THE MATH BEHIND THE MUSIC

Mathematics has been used for centuries to describe, analyze, and create music. In this book, Leon Harkleroad explores the math-related aspects of music from its acoustical bases to compositional techniques to music criticism, touching on

- overtones, scales, and tuning systems
- the musical dice games attributed to Mozart and Haydn
- the several-hundred-year-old style of bell-playing known as ringing the changes
- the twelve-tone school of composition that strongly influenced music throughout the twentieth century

and many other topics involving mathematical ideas from probability theory to Fourier series to group theory. He also relates some cautionary tales of misguided attempts to mix music and mathematics.

Both the mathematical and the musical concepts are described in an elementary way, making the book accessible to general readers as well as to mathematicians and musicians of all levels. The book is accompanied by an audio CD of musical examples.

Leon Harkleroad gives frequent lectures on mathematics and music for the Mathematical Association of America and other groups. He has been published in many journals, and he has received the George Pólya Award from the MAA for one of his papers.

OUTLOOKS

PUBLISHED BY CAMBRIDGE UNIVERSITY PRESS AND THE MATHEMATICAL ASSOCIATION OF AMERICA

Mathematical content is not confined to mathematics. Eugene Wigner noted the unreasonable effectiveness of mathematics in the physical sciences. Deep mathematical structures also exist in areas as diverse as genetics and art, finance and music. The discovery of these mathematical structures has in turn inspired new questions within pure mathematics.

In the *Outlooks* series, the interplay between mathematics and other disciplines is explored. Authors reveal mathematical content, limitations, and new questions arising from this interplay, providing a provocative and novel view for mathematicians, and for others an advertisement for the mathematical outlook.

THE MATH BEHIND
THE MUSIC

LEON HARKLEROAD

CAMBRIDGE
UNIVERSITY PRESS

CAMBRIDGE UNIVERSITY PRESS
Cambridge, New York, Melbourne, Madrid, Cape Town, Singapore, São Paulo, Delhi

Cambridge University Press
32 Avenue of the Americas, New York, NY 10013-2473, USA

www.cambridge.org
Information on this title: www.cambridge.org/9780521009355

First published 2006
Reprinted 2007 (twice), 2008, 2009

Printed in the United States of America

A catalog record for this publication is available from the British Library.

Library of Congress Cataloging in Publication Data

Harkleroad, Leon, 1955–
The music behind the music / Leon Harkleroad.
  p.   cm. – (Outlooks)
Includes index.
ISBN-13: 978-0-521-81095-1 (hardback)
ISBN-10: 0-521-81095-7 (hardback)
ISBN-13: 978-0-521-00935-5 (pbk.)
ISBN-10: 0-521-00935-5 (pbk.)
1. Music in mathematics education.   2. Mathematics – Study and teaching.
3. Music theory – Mathematics.   4. Music – Acoustics and physics.   5. Pattern
formation (Physical sciences)   6. Composition (music)   I. Title.   II. Series.
QA19.M87H37   2006
510–dc122   2006014720

ISBN 978-0-521-81095-1 hardback
ISBN 978-0-521-00935-5 paperback

For Cynthia

Nagyon szeretlek, legkedvesebbem!

CONTENTS

PREFACE

For centuries people have explored connections between mathematics and music. Yet, when I started writing this volume, very little in the way of books surveyed the range of these connections. You could find monographs devoted to one specific niche or another, as well as any number of journal articles. But you would have had a hard time locating a reputable source that gathered several of the important topics in one place and gave them a broadly accessible presentation. Fortunately, that situation has been changing recently. I hope that this book helps with such change. In Chapter 1, I give an overview of the contents of the subsequent chapters. I will just say here that I have tried for a blend of standard, core topics with less familiar ones. In particular, I have highlighted some recent work on math and music that may be new even to aficionados of the subject.

The potential readership for an interdisciplinary book like this encompasses a wide variety of backgrounds. Accordingly, I have aimed to presume as little as possible on your prior experience in either mathematics or music. The book, in part, grew out of lectures I gave in a course that had no prerequisites. So, the students ranged from math majors who were accomplished musicians to math-phobes who could not read or perform music at all. I don't start totally from scratch in these pages as I did in the course, but I have written with a fairly general audience in mind. Again, those already familiar with math and/or music should still find much new material here. Despite the book's origins, I have not written it as a textbook. However, anyone inclined to take on the highly enjoyable task of teaching a math-and-music course could certainly make use of this in conjunction with the class.

ACKNOWLEDGMENTS

This book developed in tandem with three other math-and-music projects of mine: the course at Cornell University that I previously mentioned, a seminar that I organized at Cornell, and a workshop that I have offered under the auspices of the Mathematical Association of America. Thanks to Cornell and the MAA for letting me organize these activities and especially to all the participants for sharing an interest in math and music.

Particular thanks to Kevin Hamlen for allowing me to present his previously unpublished work here.

A huge thank you to concert musician and audio engineer Graybert Beacham for the time and skill he devoted to producing the CD. He did a masterful job of tracking down sources, recording my performances, playing the violin, and compiling it all.

Many thanks to everyone from Cambridge University Press and the MAA who helped bring this book into existence: Pooja Jain, Katie Hew, Lara Zoble, Caitlin Doggart, Roger Astley, Don Albers, John Barrow, and Don Zagier. I must single out Jessica Farris, who did some extraordinary work in obtaining reprint permissions; David Tranah, who proposed the book in the first place; and my editor Lauren Cowles, whose helpful advice and abundant patience I very much appreciate.

ACKNOWLEDGMENTS

Above all, my utmost gratitude to my wife Cynthia for proofreading, suggestions, encouragement, and support. This book owes a great deal to her – but she specifically disclaims responsibility for any of my puns contained in it.

ONE

MATHEMATICS AND MUSIC, A DUET

When I mention to someone my work on a book about mathematics and music, I tend to get one of two responses:

"Yeah, they really are a lot alike, aren't they?"

"What in the world could the one have to do with the other??"

Math and music *do* have much in common. At heart, abstract patterns form the stock-in-trade of both. To express these patterns, each field has developed its own symbolic language, used the world over regardless of nationality. And the two areas, although in different ways, combine the intellectual and the aesthetic in a wonderful blend. Unfortunately, nonmusicians often remain unaware of the rich intellectual content of music, and nonmathematicians likewise of the equally rich aesthetic side of math. That accounts for many of the skeptical reactions I mentioned above. An anecdote of Raymond Smullyan provides a different slant on these misperceptions. In his book, *5000 B.C.*, Smullyan told of a mathematician hearing people talk about connections between math and music. The mathematician, after looking puzzled, blurted out, "But I don't see the likeness; after all, mathematics is beautiful!" Such apocryphal stories notwithstanding, math and music do appeal to their practitioners in similar ways. To quote the nineteenth-century mathematician J. J. Sylvester, "May not Music be described as the Mathematic

1

of sense, Mathematic as Music of the reason? the soul of each the same!"

But beyond the generalities of shared traits, the two areas have many specific, direct links. Over the years, people have used math to describe, analyze, and create music. This book constitutes an informal smorgasbord of several such applications to music.

The Good Old Summary Time

One of the more obvious contacts between the two fields is through the middleman of physics. The acoustical side of music lends itself well to mathematical analysis. Chapter 2 lays out some of the basic mathematical relationships behind musical pitch, timbre, and overtones. In particular, I examine why some notes harmonize well with others. Attempts to incorporate such pleasing harmonies have led to various schemes for tuning musical scales. The third chapter compares some of these tuning systems.

These acoustical considerations deal with the raw material of music. In the visual arts, the parallel study would be the physics of light and color. But with visual art, mathematics also enters into the compositional aspects. Think of the theory of perspective drawing or the works of M. C. Escher. Similarly, math plays a role in musical composition. Sometimes composers have explicitly employed mathematical techniques in their works. At other times composers have built structures that, although designed by purely musical considerations, can profitably be described in mathematical terms. The math may not present itself as obviously as the geometry in a painting, but it can be just as vital.

The fourth and fifth chapters both draw on the mathematical area of group theory, though in different ways. The former chapter inspects the group theory behind some standard tricks of the composer's trade, such as transposing a melody or playing it backwards. In Chapter 5, I concentrate on the old English tradition of ringing the changes on tower bells. Change-ringers had actually used many group-theoretic ideas in this specific context long before mathematicians developed the subject. For the next chapter the focus shifts to randomness and probability as used by various composers. From musical dice in Mozart's time to recent computer work, musicians have been intrigued by the artistic possibilities of chance. Chapter 7 investigates some structural qualities of music. Specifically, I discuss how the same patterns can appear at different levels within a piece. Earlier I compared music with visual art. Of course, the two sometimes interact, and math can help bind them together. Chapter 8 details various mathematical strategies for building music out of pictures. We will also take a look at dancing, where group theory returns to the scene.

Given all of the areas of overlap between math and music, it can be easy to get carried away and start seeing connections that aren't necessarily there. This danger looms especially large when a mathematical analysis is imposed on a musical composition after the fact. The final chapter contains some cautionary tales of dubious attempts to meld the two fields.

Mathematics and music go back a long way together. In the Western tradition, their partnership traces back at least 2500 years, when the Pythagoreans explored the math–music connection. The seven liberal arts of the Middle Ages consisted of the quadrivium, the path to knowledge, and the trivium, the path to eloquence. As I suggested at the beginning, many people these days associate the nature of music with the latter more than with the former. But to medieval scholars, music enjoyed good standing as a member of the quadrivium. Its mates there? Arithmetic, geometry, and astronomy. The interplay between math and music continues at present. Indeed, recent years have seen a proliferation of conference sessions and even entire conferences devoted to the links that connect mathematics and music. I hope that this book provides at least a taste of the fascinating results that occur when math and music combine forces.

TWO

PITCH: THE GROUND OF MUSIC

What makes a note a note? Certainly, music often features sounds that aren't notes with specific pitches. From a marching band's snare drums to flamenco castanets to the cannons in Tchaikovsky's *1812 Overture*, music can make much use of nonpitched sounds. But usually when we think of music, we think about tones at various pitches, some higher, some lower. Both melody and harmony depend on such tones. But how do they really differ from other sounds? The key lies in repetition – very rapid repetition.

Window Screens and Siren Whistles

Whether they have pitches or not, the sounds we hear arise from variations in air pressure. For example, the graph in Figure 1 depicts the "white-noise" sound of a not-tuned-in television set. More precisely, the graph represents the kind of changes in air pressure that we would perceive as sound while watching the television "snow." Reading Figure 1 horizontally from left to right tracks a certain time span.

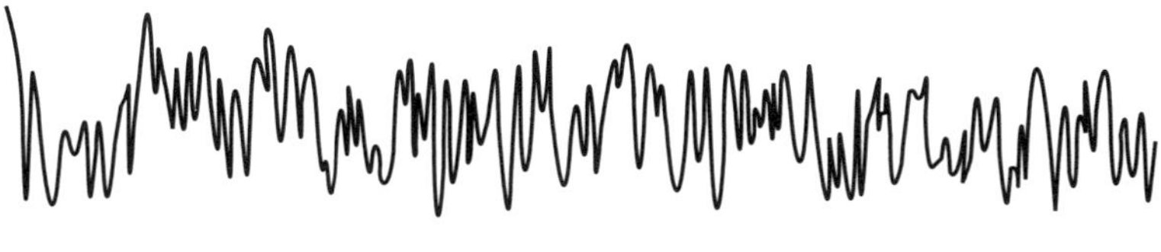

Fig. 1

5

Fig. 2

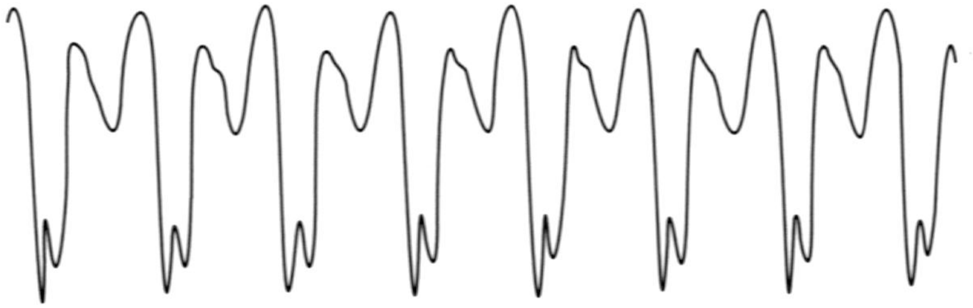

Over that span the air pressure, represented on the vertical scale, varies. The greater the variation, the louder the sound. In fact, people can respond to sounds due to deviations as little as 0.00000002 percent of the baseline pressure!

Now contrast that graph with the one in Figure 2, depicting the sound of a clarinet. The difference is immediately apparent. On the one hand, the white-noise graph jitters up and down without any noticeable pattern. On the other hand, the clarinet graph clearly consists of a certain shape repeated over and over. That regularity is precisely what gives the sound a pitch, making it a note. In this case the pattern lasts about $1/750$ of a second, so it repeats about 750 times per second. Customarily, we say that it repeats at a rate of about 750 *Hertz*, often abbreviated as 750 Hz.

In general, the rate at which an air pressure pattern recurs – usually referred to as the *frequency* – determines how high or low the corresponding pitch will be. So-called Middle C, a note of medium pitch, corresponds to a frequency of around 262 Hz. Increasing the frequency, that is, making the pattern repeat faster, raises the pitch (but see Sidebar 1).

You can observe this phenomenon by running your fingernail over a window screen. Because of the screen's grid, you will induce a somewhat regularly repeated pattern of vibration in the surrounding air. This leads to a sort of a tone. The screen possesses enough nonuniformity that the sound won't be very musical, but as you scrape the screen more rapidly, you can expect to hear the "tone"

For simplicity, I shall treat the pitch of a note as being determined by the frequency of air-pressure variations. The true state of affairs is more complex. By definition, pitch is a subjective sensation in response to a note. Factors other than frequency can influence the pitch we hear. For instance, when the loudness of a note changes, some people may sense a change in pitch as well, even if the frequency remains constant.

Vision presents a similar situation. Various factors, from optical illusions to medical conditions, influence color perception, but still certain frequencies of light waves are said to correspond to certain colors.

SIDEBAR 1

get higher. Using a comb or a zipper, you can produce similar (and less grating!) effects.

For something a bit more melodious, pick up a little siren whistle (Figure 3) from a toy store. When you blow into the siren, a pinwheel blade spins around. The blade is punctured with holes, as is a stationary plate that lies behind it. Air will escape through the whistle's far end at those moments when the holes in the blade line up with the stationary holes. So as you blow, there results a periodic stream of puffs of air, creating a tone. The harder you blow, the faster the blade spins and the more frequently puffs recur. Consequently, the pitch rises. Thus, you can control the notes and play melodies simply by

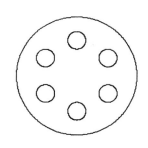

Fig. 3

Side view   End view

The siren whistle is an adaptation of the much larger, hand-cranked siren. (Of course, modern emergency vehicles produce siren sounds electronically.) The siren takes its name from the mythological sea-nymphs who lured sailors by irresistibly dulcet singing. Now a lot can be said about the tonal quality of an instrumental siren, but "dulcet" surely doesn't make the list! Rather, the name comes from the instrument's ability to produce sounds underwater. Several noted composers of the twentieth century – Satie, Hindemith, Varèse, to name some – occasionally used sirens in their compositions. So has Peter Schickele in his tongue-in-cheek P.D.Q. Bach music, with sirens attached to trombones and to bicycles.

adjusting the strength with which you blow into the siren. Track 1 of the CD contains a demonstration of this.

Same Difference

Now for another experiment you can try. This one doesn't require any equipment at all, except a bunch of cooperative friends. Sing a note and ask your friends to sing that note along with you. They may hesitate some, depending on how shy or not they are about singing in front of others. But at least when I have performed this experiment, people don't bat an eye at the concept of everyone singing the same note, and they agree when the group has reached more or less of a unison. There's just one catch. Unless everyone's voice falls in the same range, people will be singing notes of quite different pitches. In particular, as a rule there will be a high pitch sung by some of the women, a low pitch sung by some of the men, and a pitch or two in between sung by both men and women. Still we have no trouble thinking of these different frequencies as somehow yielding the same note. And this does not just arise as an artifact of our cultural conditioning. Around the world, many a culture has a notion of unison singing that somehow equates different pitches – the very pitches that we equate.

By the end of this chapter, we will see some reasons for why we respond to certain pitches this way. For now, the pertinent fact is that such related tones possess frequencies that are numerically related in a very simple fashion. *Doubling the frequency of a note results in a higher note that we nevertheless perceive as, in some sense, the same as the original.* In musical terminology, the note with double the frequency lies an *octave* higher. Going up another octave creates another note in that "family," at four times the frequency of the starting pitch. Likewise, descending by octaves – halving the frequency each time – produces lower notes in that family, or as it is sometimes called, *pitch class.*

To identify and lump together different, yet related things – such as notes separated by octaves – is typical of human thought in general. At an early age we learn that certain entities, although of widely varying sizes, shapes, and colors, all bark, wag their tails, etc., and we classify them together as the same animals, namely dogs. Such identifications occur frequently in mathematics. In high-school geometry, the congruence of triangles furnishes an example. Triangles are somehow equated, even if they possess different locations and orientations, as long as they have the same size and shape. The notation $\simeq$ used for congruence tweaks the equals sign to suggest this sort-of-sameness. A simpler example, actually very much in the same vein as pitch classes, is that of classifying numbers as odds and evens. Whenever a collection of mathematical objects is broken up into clusters of objects lumped together, each such cluster is known as an *equivalence class.* Objects in the same equivalence class are called *equivalent,* and often a notation like $\simeq$ denotes the equivalence relation between those objects.

SIDEBAR 3

Fig. 4

You can easily recognize notes of the same pitch class on a piano keyboard (Figure 4). The black and white keys come in a 12-key pattern that repeats up and down the piano. Notes that belong to the same pitch class, such as those marked in the figure, have the same relative position in the pattern and thus lie 12 or 24 or 36, and so forth, keys apart. In Figure 4, each marked note is called A (the notes of a pitch class all have the same name.) The lowest A corresponds to a frequency of 27.5 Hz, since its pattern of air-pressure variation repeats 27.5 times every second. The highest A, seven octaves up, has a frequency of 27.5 doubled seven times – that is, $27.5 \times 2^7 = 3{,}520$ Hz. For comparison, people typically can hear pitches from 20 to 20,000 Hz.

Pythagoras' Fifth

The ancient Greeks – specifically, the Pythagoreans – discovered that octaves related to doubling and halving. Rather than in the context of frequencies, though, the Greeks noted the connection in terms of the lengths of strings in stringed instruments. Players of instruments like lyres and so on had long known techniques for obtaining different pitches. The Pythagoreans quantified this. Say that you have a string whose ends are fixed in place. Plucking the string produces a certain note. Now, if you keep the center of the string fixed and pluck one half, you get a note an octave higher.

Interestingly, more than 2000 years elapsed before people realized how frequency fit in. The French mathematician and priest Marin Mersenne (1588–1648) saw that the crux of pitch lies in frequency, and in 1636, he published an equation that exhibited how the vibration frequency of a string depends on the string's length.

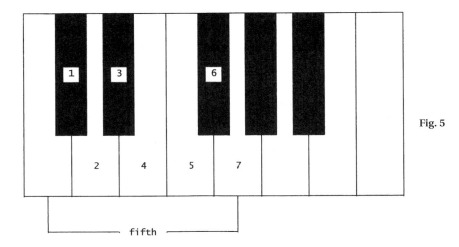

Fig. 5

(Other factors come into play, too. As anyone who has ever tuned a stringed instrument knows, the string's tension definitely influences the pitch!) Around that time, Galileo (1564–1642) independently did work similar to Mersenne's.

Besides figuring out what makes octaves tick, the Pythagoreans made similar discoveries about other pairs of musically related notes. Take, for instance, the two "twinkle"s in *Twinkle, Twinkle, Little Star*. Those two notes not only work well in succession, they also blend harmoniously when sounded together. Unlike notes an octave apart, we do not identify them with each other. Still, they are definitely connected. We describe those notes as being a *fifth* apart, and the interval of a fifth ranks second only to the octave in musical importance. On a piano keyboard, a fifth corresponds – more or less – to a distance of seven keys (Figure 5). The interval between successive piano keys is called a *semitone*, so a fifth more or less spans seven semitones.

All this raises several questions.

First, why is the interval called a fifth? Because the note lying a fifth up from a starting note is the fifth note in the major scale based on that starting note. For example, the C major scale consists of the white keys on the piano, and in Figure 5, G is the fifth white key, counting from C as the first.

Second, why did I add the qualification "more or less" to my statement that a fifth corresponds to seven semitones? The reason is a whole story in itself. In the next chapter we will explore this matter fully.

Finally, does the interval of a fifth, like that of an octave, result from a simple frequency relationship? Yes, indeed. The Pythagoreans, as mentioned earlier, talked about string lengths rather than frequencies. But in effect, they found that multiplying a note's frequency by 3/2 raises that note by a fifth.

Another important interval, the *fourth*, complements the fifth. Start at C, for instance, and go up a fifth to G and then up to the next C (Figure 6). (These notes, played on the trumpet, begin Richard Strauss' *Also Sprach Zarathustra*, which was used to accompany the opening scene of the movie *2001: A Space Odyssey*.) The interval from the G to the C above it constitutes a fourth. Referring to Figure 6, we can see that this interval corresponds to a frequency ratio of $\frac{2f}{\frac{3}{2}f} = 4/3$. In other words, multiplying the frequency of a note by 4/3 raises it by a fourth. Figure 6 also shows that (subject to the "more

Fig. 6
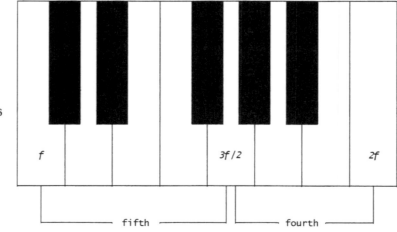

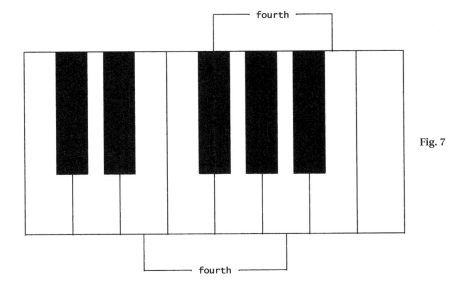

Fig. 7

or less" caveat) a fourth comprises a five-semitone stretch. Thus, the intervals in Figure 7 also form fourths.

Two other intervals will show up in the pages ahead. At the start of *When the Saints Go Marching In*, both "Oh, when" and "go mar-" leap by a *major third*, with a frequency ratio of 5/4. For convenience, I shall refer to such an interval simply as a *third.* Likewise, the first two notes of *My Bonnie Lies Over the Ocean* make up a *major sixth* (again, just *sixth*, for short) with a frequency ratio of 5/3. As Figure 8 indicates, a third and a sixth span four and nine semitones, respectively.

Sine Is the Bottom Line

With octaves, we have the phenomenon of different pitches that somehow sound the same. Let's now examine how notes of the same pitch can sound different. This opposite consideration, besides being significant in its own right, surprisingly also sheds light on why we respond to octaves and other intervals the way we do.

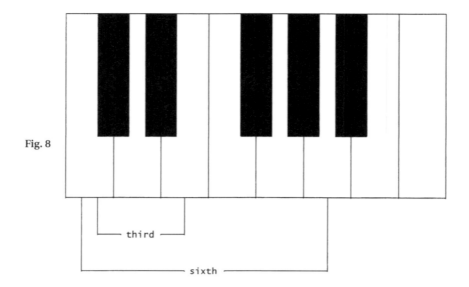

Fig. 8

One obvious way to change a tone, while preserving its pitch, is to alter its volume. As I noted earlier, the volume correlates with the magnitude of variations in air pressure. In terms of the graph of air pressure versus time, you louden a note by stretching vertically (Figure 9).

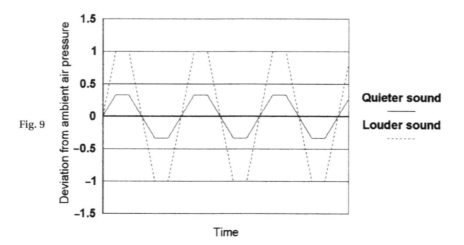

Fig. 9

14

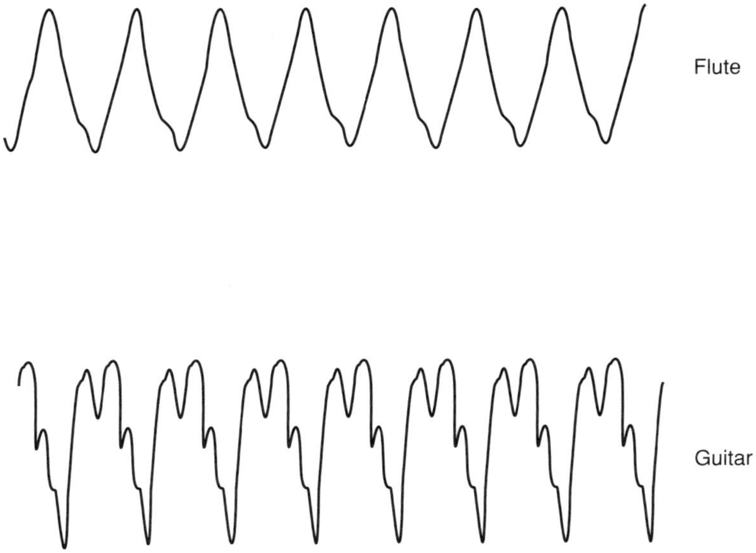

Flute

Fig. 10

Guitar

But that's not the only way to change a tone. Two notes of the same pitch and the same volume can still sound quite different if they come from different instruments. Figure 10 displays graphs corresponding to a flute and a guitar. Aside from any considerations of pitch or volume, the two graphs clearly set themselves apart by their shapes. Again, when you repeat a pressure pattern at a certain frequency, you get a note at a certain pitch. The nature of the repeated pattern itself determines the note's tonal character (in musical terminology, *timbre*), distinguishing a flute from a guitar from a French horn.

At first glance the infinite variety of possible shapes might seem to make timbre intractable to analyze. Fortunately, some structure comes to the rescue. Every tone can be obtained by combining certain simple building blocks, called *pure tones*. What do I mean by "combine"? Musically, just to play the tones simultaneously. In terms of the air-pressure graphs, for each moment of time you

SIDEBAR 4

Like pitch (see Sidebar 1), timbre is subject to considerations of subjective perception. Beyond that, other complications arise. For one, a given instrument can produce a variety of pressure patterns and thus of timbres, according to where in the instrument's range a note lies, how loud the note is, and so on. Moreover, when a player sounds a note, the pressure variations go through a complex initial phase before settling into a regular pattern. This phenomenon, known musically as *attack*, certainly contributes to the distinctive tonal characters of different instruments.

put together the respective deviations from the baseline pressure. Figure 11 illustrates how merging two such graphs creates a more complicated one – and thereby a new timbre.

The graphs in Figures 11a and 11b possess the shape characteristic of pure tones. Graphs with that shape are known as *sine curves*. In trigonometry, the scene of most people's first contact with sines, they show up as ratios of lengths of sides of right triangles. A nonmathematician might well wonder why sines would appear in the present context. But besides their trigonometric properties, sine curves also have the feature that they can generate graphs as complex as those in Figures 2 and 10. From the musical viewpoint: *you can construct a note with a frequency f and any timbre you want by blending pure tones whose frequencies are multiples of f, that is, 2 f, 3 f, and so forth.* You obtain the various timbres at frequency f by adjusting the relative volumes of these component pure tones. Track 2 of the CD was produced by recording a guitar note. The pure tones forming it were then electronically separated out. In the track these components are put back together, one at a time, until the full guitar sound appears. Electronic synthesizers work in a similar fashion. Circuitry can easily generate pure tones. To mimic different instruments, a synthesizer just mixes pure tones in appropriate proportions.

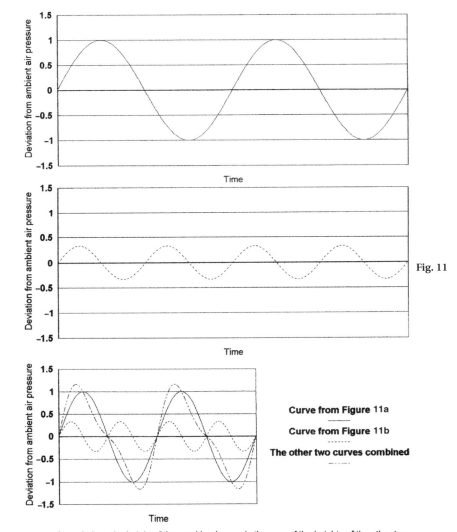

Fig. 11

At each time, the height of the combined curve is the sum of the heights of the other two.

Now it's the mathematician's turn to wonder. For there are other collections of curves with generative properties similar to those of the sines. Do we use sines simply for their mathematically nice characteristics, or do pure tones really play a distinguished role in our

17

Overtones have been observed for a long time. In the fourth century BC, Aristotle asked, "Why does the low note contain the sound of the high note? . . . Why is it that in the octave, the concord of the upper note exists in the lower, but not vice versa?" Mersenne also dealt with overtones in his groundbreaking research on vibrations. He discussed "why a vibrating string gives several sounds simultaneously . . . Aristotle did not know that the struck string gives at least five different sounds simultaneously . . . These sounds follow the ratio of the numbers 1, 2, 3, 4, 5." Then in 1822, came the big breakthrough, due to Jean Baptiste Joseph Fourier (1768–1830). Fourier discovered that, in general, a repeating wave pattern can be broken up into (possibly infinitely many) sines in the way that we have discussed. He did so in the context, not of sound, but of heat conduction. Following this, Georg Ohm (1787–1854), most famous for his work with electricity, proposed that the ear breaks down notes into their component pure tones. The physicist, anatomist, and physiologist Hermann von Helmholtz (1821–1894) built upon Ohm's idea, besides making many other important contributions to the physics of music. Our present understanding of how the ear processes notes owes much to the Nobel-Prize-winning research of Georg von Békésy (1899–1972).

hearing? In fact, the human ear does function on the basis of pure tones. Say that you hear a clarinet note (Figure 2). The basilar membrane in your ear will process the sound by responding to the separate pure-tone components that constitute the clarinet note. The greater the relative volume of a component, the more strongly the membrane will respond to that component. Your perceptual system then gathers these pure-tone responses and assimilates them into your sensation of a clarinet note. Curiously, although our ears do it automatically for us, breaking up a note into pure tones is much more sophisticated mathematically than blending tones. We have seen, in Figure 11, how easily curves can be combined – it's nothing more than adding and subtracting. On the other hand, to reconstruct the original sine curves from the combination calls for integral calculus, as justified by so-called Fourier analysis. Not bad for a basilar membrane . . .

Overtones Underpinning Intervals

With this background, let's revisit octaves and other intervals. We have seen that a note of frequency f also contains, smuggled within it, pure tones of higher frequency. These higher-frequency components are often called *overtones* of the original note. In particular, the pure tone an octave up, at $2f$, is an overtone. As a general (though often violated) rule, the relative volumes of overtones decrease as their frequencies increase, so among the overtones, the one at $2f$ tends to be prominent. Furthermore, all of the overtones of a note at $2f$ (i.e., $4f, 6f, 8f$, etc.) are overtones of the note at f. No wonder, then, that we sense a close connection between notes an octave apart! You can demonstrate this relationship for yourself at a piano. Take any two notes an octave apart. Hold down the key for the upper one, which leaves the string free to vibrate. Then strike, but don't hold down, the key for the lower one. This lower note plays and quickly dies out. But its $2f$ overtone, at the right frequency for the higher note, sets that string in motion, and you will hear the higher note long after the lower has faded.

What about the interval of a fifth? A note at $\frac{3}{2}f$ is not an overtone of a note at f, but it lies an octave below f's overtone at $3f$. Of course, $3f$ also occurs as an overtone to $\frac{3}{2}f$. Indeed, every other overtone of $\frac{3}{2}f$ is also an overtone of f. So, although notes at $\frac{3}{2}f$ and f do not have enough in common for us to equate them as we do $2f$ and f, they relate to each other enough to sound harmonious together. Similar considerations apply to the other intervals we have discussed.

Unfortunately, frequency ratios like 3/2 and 5/3 represent ideals that are unachievable even at the theoretical level. Or rather, any one such ratio can be managed, but taken together they are incompatible. For example, you cannot make all octaves 2/1 and all fifths 3/2 at the same time. In the next chapter, I will explain why, as well as discuss various musical compromises people have devised to deal with this situation.

THREE

TUNING UP

We saw in the previous chapter that pleasant harmonies result from notes whose frequencies are related by simple ratios such as 2 (the octave) and 3/2 (the fifth). But there's a catch. Ultimately, these ratios are incompatible with each other, so that your musical intervals cannot all simultaneously live up to the ideal. You must make compromises. Which compromises you do and don't make will depend on which musical features you consider more or less expendable. In this chapter, we will look at why the intervals cannot all achieve perfection simultaneously and at various ways of coping with this problem.

You Take the High Road, and I'll Take the Low Road

Let's start with a very simple incompatibility. For convenience, Figure 1 summarizes the basic intervals as I have described them. Take a note of frequency f. If you go up by a fourth, you get a note of frequency $\frac{4}{3}f$. Now go up by a sixth from that note. Your new frequency will be $\frac{5}{3} \times \frac{4}{3}f = \frac{20}{9}f$. Having risen by a fourth and then a sixth, you end up $5 + 9 = 14$ semitones higher than you began. However, you can climb fourteen semitones another way, namely, by two consecutive fifths, each spanning seven semitones. The first fifth puts you at a note of frequency $\frac{3}{2}f$, and a fifth up from

Fig. 1

Interval	Frequency ratio	Semitones
Third	5/4	4
Fourth	4/3	5
Fifth	3/2	7
Sixth	5/3	9
Octave	2/1	12

Note: I shall refer to intervals with these exact frequency ratios as *ideal* or *perfect*, although this use of the word "perfect" is slightly at odds with standard musical terminology.

SIDEBAR 1

Consider two pure tones of frequencies 300 and 330 Hz. What happens when they occur simultaneously? Recall from the last chapter that we can combine their respective graphs to obtain the graph of the joint sound. It turns out as follows.

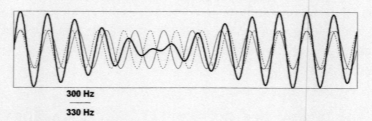

300 Hz
⎯⎯⎯
330 Hz
- - - - -
The other two curves combined
⎯⎯

The combination's basic pattern is that of a sine curve of frequency 315 Hz, but with the curve's height fluctuating at a frequency of 15 Hz. Since the height of the curve determines the volume of the sound, this note will fade in and out, producing a somewhat disagreeable pulsating effect. Instrumentalists and tuners listen for such pulses when trying to match an instrument's tuning with that of another instrument or a tuning fork. With proper matching, no pulsing occurs.

there lands you at a frequency of $\frac{3}{2} \times \frac{3}{2} f = \frac{9}{4} f$. Supposedly, the two routes end at the same place, fourteen piano keys higher than where you began, but they actually yield different frequencies: $\frac{20}{9} f$ versus $\frac{9}{4} f$.

Now admittedly the discrepancy between 20/9 ($= 2.222\ldots$) and 9/4 ($= 2.25$) is not very great. As a result, the frequencies will lie fairly close to each other. That poses its own problems, though. Notes that are close, but different, tend to clash when sounded simultaneously. Just play two adjacent piano keys at the same time for a slightly nerve-grating experience. (See Sidebar 1 for a discussion of why such notes sound unpleasant together.) So, if in a duet, two singers started on the same note, and the first singer went up a perfect fourth and then a perfect sixth, while the other singer went up two successive perfect fifths, they would end up on slightly different notes and sound uncomfortably out of tune.

Something, then, has to give. As we have previously seen, the fifth is a very important interval related to the $3f$-overtone, while the fourth and fifth together comprise an octave. Thus, we could reasonably choose to make the fifth and fourth have their ideal values and let the sixth give way a little. In other words, we can assign frequencies as in Figure 2, with a perfect fourth from C to F and perfect fifths from C to G and from G to D. The sixth from F to D here has a frequency ratio of 27/16 ($= 1.6875$), rather than the perfect 5/3 ($= 1.666\ldots$).

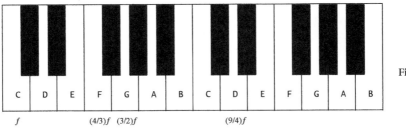

Fig. 2

We certainly want to preserve the frequency ratio of 2 for the octave, so we can fill in some more frequencies as follows (see Figure 3).

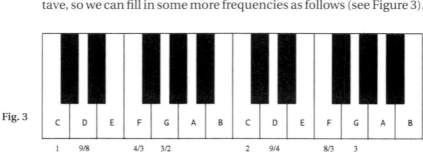

Fig. 3

Note: For simplicity, I have omitted the factor of f from each frequency.

Notice that because we have used perfect octaves and fifths among the C's, D's, and G's, we automatically obtain a perfect fourth between each D and its succeeding G: $\frac{\frac{3}{2}f}{\frac{9}{8}f}$ and $\frac{3f}{\frac{9}{4}f}$ both equal 4/3. We can now similarly opt for a perfect fifth from D to A (Figure 4)

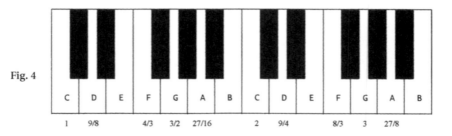

Fig. 4

and then from A to E (Figure 5)

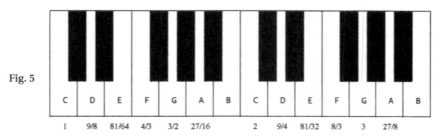

Fig. 5

24

and finally from E to B (Figure 6).

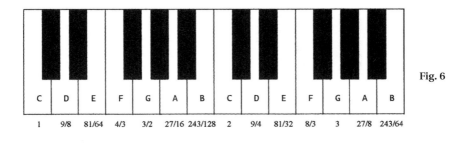

Fig. 6

At this point we have assigned frequencies to all the white keys in such a way that all octaves, fifths, and fourths among these notes meet the ideal specifications. All sixths (C–A, D–B, F–D, and G–E) correspond to ratios of 27/16. Likewise, all thirds have ratios of 81/64 (= 1.265625), as opposed to the perfect 5/4 (= 1.25). If fourths and fifths figure more prominently in the music than thirds and sixths, this tuning scheme can work well, ensuring that the favored intervals sound harmonious and shunting the discrepancies off to the less-used intervals. In fact, this system – known as Pythagorean tuning – dates back to the ancient Greeks, whose music made much use of fourths and fifths.

A Just Song at Twilight

But changing musical styles called for other tuning systems. Music composed during the Renaissance gave increased prominence to thirds and sixths. Accordingly, musicians of that period devised tunings that incorporated perfect thirds and sixths, even if they sacrificed an occasional perfect fourth or fifth. We could, for example, start out by using the ideal ratios for all intervals built up from the starting note of C (Figure 7).

25

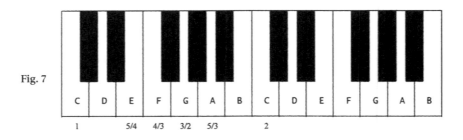

Fig. 7

Extending this to the succeeding notes, we obtain Figure 8.

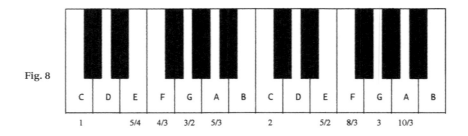

Fig. 8

Notice that the intervals between notes other than C also work out properly. For example, from E to A is a perfect fourth ($\frac{\frac{5}{3}f}{\frac{5}{4}f} = \frac{4}{3}$), and from G to E is a perfect sixth ($\frac{\frac{5}{2}f}{\frac{3}{2}f} = \frac{5}{3}$). We can also tune the B's to achieve perfect E–B fifths and G–B thirds (Figure 9).

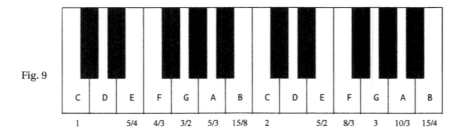

Fig. 9

At this stage, though, we must compromise. In assigning a frequency to D, we cannot make both the G–D fifth and the F–D sixth be perfect intervals. (Indeed, this is the same situation we faced in

Figure 2.) By favoring the fifth and making it perfect, we obtain the following system (Figure 10).

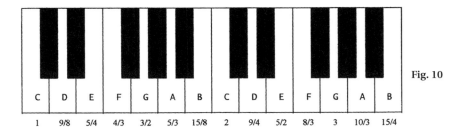

Fig. 10

C	D	E	F	G	A	B	C	D	E	F	G	A	B
1	9/8	5/4	4/3	3/2	5/3	15/8	2	9/4	5/2	8/3	3	10/3	15/4

Notice that this choice also produces an ideal D–B sixth. On the other hand, not only the F–D sixth, but also the D–A fifth, will be off. We could have ensured perfect intervals for D–A and F–D at the expense of G–D and D–B, but the scheme represented in Figure 10 works a little better musically.

This system belongs to a family of tunings called *just intonations*, which employ simpler numbers than the Pythagorean tuning and give more intervals their ideal frequency ratios. However, with this tuning we lose some consistency. Not all fifths correspond to the same ratio, D–A yielding 40/27 rather than the 3/2 of all the other fifths. Likewise, the F–D sixth possesses a ratio of 27/16, as opposed to the rest at 5/3.

Imperfectly Consistent

I may have given the impression that the dilemma of tuning comes from trying to reconcile fourths and fifths on the one hand with thirds and sixths on the other. The problems run much deeper, though. We cannot even attain compatibility of perfect octaves and fifths, the two most important intervals. True, the Pythagorean tuning we have looked at contains nothing but 2/1 octaves and 3/2 fifths. But I have fudged a bit by applying that system only to the white keys on the piano. If we start at C and pile up successive fifths, we hit every pitch

Fig. 11

class – white *and* black keys – before finally returning to a C (see Figure 11). And when we include all those fifths, we must fall short of the ideal in our tuning. As Figure 11 indicates, going up twelve successive fifths of seven semitones each should land us at the same place as going up seven successive octaves of twelve semitones each. However, twelve perfect fifths would correspond to a frequency ratio of $\left(\frac{3}{2}\right)^{12}$, while seven octaves would correspond to 2^7. These numbers are close (around 129.75 and 128, respectively), but not the same.

The octave ratio of 2/1 is usually taken to be inviolable, as firmly rooted in overtones as it is. So, that leaves us to tinker with the fifths. We can do so in two different ways. We could make as many fifths perfect as possible, absorbing the discrepancy in a fifth that we hope doesn't get used too often. Or we could give up on perfection altogether and go for consistency by tuning all fifths at the same ratio. Rather than having one particularly bad fifth, we would make all fifths just a little bit off. How much off? If each fifth shares the same ratio r, we want to pick that r so that twelve fifths exactly match seven octaves, that is, $r^{12} = 2^7$. This means $r = \sqrt[12]{2^7}$, approximately 1.4983. The deviation from the perfect value of 1.5 is too small for the normal human ear to perceive.

If every fifth corresponds to a $\sqrt[12]{2^7}$ ratio and we keep perfect ratios of 2 for the octaves, we come up with the tuning in Figure 12.

C	C♯	D	D♯	E	F	F♯	G	G♯	A	A♯	B	C
1	$\sqrt[12]{2}$	$\sqrt[12]{2^2}$	$\sqrt[12]{2^3}$	$\sqrt[12]{2^4}$	$\sqrt[12]{2^5}$	$\sqrt[12]{2^6}$	$\sqrt[12]{2^7}$	$\sqrt[12]{2^8}$	$\sqrt[12]{2^9}$	$\sqrt[12]{2^{10}}$	$\sqrt[12]{2^{11}}$	2

Each number represents the ratio of the frequency of the note to the frequency of the starting C.

Fig. 12

In the same manner as just intonations accommodated the musical styles of the Renaissance, so does the use of equal-tempering reflect later musical fashions. In particular, over the years music became more and more likely to rely on all twelve pitch classes. This trend reached a climax with the flourishing of so-called twelve-tone composers during the early and middle twentieth century (see Chapter 4). People considered equal-tempering at a theoretic level as far back as Mersenne or earlier. However, the research of Alexander John Ellis (1814–90) showed that even as late as 1885, keyboards tuned by that system were the exception, rather than the rule. Historians of music have differed in their judgments as to whether J. S. Bach (1685–1750) intended his *Well-Tempered Clavier* and other works for a keyboard that was tempered equally or otherwise.

In this scheme the frequency ratio for any semitone, that is, any pair of adjacent notes, will equal $\sqrt[12]{2}$. Therefore, we obtain total consistency; not only do all fifths share a common ratio, the other intervals behave likewise. For example, any third, since it consists of a four-semitone span, must correspond to a ratio of $\sqrt[12]{2^4}$.

This system of tuning, known as *equal-tempering*, is the standard tuning nowadays because it puts all notes and all intervals of the same type on an equal footing (see Sidebar 2). Again, it achieves its consistency by sacrificing perfect intervals. No interval other than the octave will have its ideal frequency ratio. We have seen, however, that the fifths come quite close. Consequently, fourths will also be very near to perfect. Thirds and sixths, on the other hand, fare less well. To examine those intervals and the often small differences between equal-tempering and the other tuning systems, we will measure intervals in a more convenient manner appropriate to such subtle distinctions.

C	C♯	D	D♯	E	F	F♯	G	G♯	A	A♯	B	C
0	100	200	300	400	500	600	700	800	900	1000	1100	1200

Each number represents how many cents the note lies above the starting C.

Fig. 13

Making Cents

Musical theorists and instrument tuners customarily break up an octave into 1,200 equal, tiny steps. This means that with equal-tempered notes a semitone consists of 100 such steps, which are called *cents*. An equal-tempered fifth thus corresponds to a difference of 700¢, a third to 400¢, and so on. Figure 13, in terms of cents, certainly looks nicer than Figure 12, which presents the same information in terms of frequency ratios. And with the cents system, calculations become simpler, too. See Sidebar 3 for more of the mathematics underlying cents.

A perfect fifth corresponds to a difference of about 702¢. As I mentioned earlier, the two-cents discrepancy between that and the equal-tempered 700¢ lies below the level of noticeability. On the other hand, a perfect third corresponds to a difference of about 386¢. Sharp ears can detect that an equal-tempered third of 400¢ is too wide an interval. An equal-tempered sixth stretches even more, 16¢ wider than a perfect sixth.

	C	D	E	F	G	A	B	C
Equal	0	200	400	500	700	900	1100	1200
Pythagorean	0	204	408	498	702	906	1110	1200
Just	0	204	386	498	702	884	1088	1200

Each number represents how many cents the note lies above the starting C.

Fig. 14

Since an octave corresponds to a frequency ratio of 2, an interval of n octaves corresponds to a ratio of 2^n. Given a frequency ratio r, then, the corresponding number of octaves is the value of n that makes $2^n = r$. In mathematical symbolism, $n = \log_2 r$, the base-2 logarithm of r. An octave consists of 1200¢, so the ratio r corresponds to $1200 \log_2 r$ cents.

Because of the properties of logarithms, when working with cents, one adds and subtracts, as opposed to the multiplications and divisions involved when working with frequency ratios. For example, in the Pythagorean tuning, a third (408¢) followed by a sixth (906¢) forms an interval of $408 + 906 = 1314$¢. In terms of frequency ratios, the relevant calculation would be $81/64 \times 27/16 = 2187/1024$. As I mentioned in the main text, musical practitioners tend to measure intervals via cents rather than via frequency ratios because of the much greater convenience that cents offer. Surprisingly, cents appeared on the scene fairly recently. People have been devising tuning systems for millennia, and mathematicians have worked with logarithms for centuries. But cents were just introduced in the 1800s by Alexander John Ellis, who appeared in the preceding sidebar.

Ellis was both a mathematician and a musician, although his main work focused on phonetics – his *magnum opus* was entitled *On Early English Pronunciation, with special reference to Shakespeare and Chaucer*. He translated *On the Sensations of Tone* by Helmholtz (see Sidebar 5 of the previous chapter) into English, revising and supplementing it according to his own acoustical studies. Ellis made a detailed investigation of the tuning practices of his day, aided by a "tonometer" device that consisted of 105 tuning forks!

The pluses and minuses of various tuning systems show up clearly when we display the systems in cents (Figure 14). Once more, equal-tempered fourths and fifths do quite well, but thirds and sixths run too wide. The Pythagorean tuning (Figure 6) makes all fourths and fifths perfect. Unfortunately, all Pythagorean thirds and sixths

run a quite detectable 22¢ beyond their ideal values. The just intonation of Figure 10 works fine for most intervals, but the fifth from
D to A comes short of perfection by 22¢, a definite flaw for such an
important interval. Notice also that the two-step intervals D–E and
G–A are 182¢, while C–D, F–G, and A–B are 204¢. That makes about a
ten-percent difference in width between intervals that would come
out the same in equal-tempering.

Again, the musical virtues of various tuning systems will depend
on the piece of music and the intervals it features. As an example,
Tracks 3, 4, and 5 on the CD contain the same composition played in
Pythagorean tuning, just intonation, and equal-tempering, respectively. These renditions appear courtesy of Erich Neuwirth. If you visit
his Web site (see Bibliography), you will find other pieces of music for
which the relative merits of the tuning systems seem quite different.

Throughout all this discussion of tunings, I have taken for granted
that we have twelve notes within an octave. Over the years people
have experimented with many other divisions of the octave as well.
The development of the twelve-note partition was doubtless influenced by considerations of intervals – in particular, the fact that
twelve perfect fifths closely approximate seven octaves. However,
other combinations of perfect fifths and octaves come even closer,
such as fifty-three fifths and thirty-one octaves. And indeed the
Pythagoreans did explore a fifty-three-note scale. Dividing the octave into nineteen or thirty-one steps has also attracted attention.
Our familiar twelve-note system, though, has held up well. Without
presenting the overwhelming set of options to a composer or the
huge technical challenges to a performer that fifty-three notes would,
our twelve still allow for a great range of possibilities that have yet to
be exhausted.

FOUR

HOW TO VARY A THEME MATHEMATICALLY

Musicians often take a melody, or even just a short sequence of notes, and work it over. By tweaking and altering a tune, they explore its musical qualities and can create rich results out of very simple material. A jazz player improvises around a Gershwin tune. A conservatory composer methodically constructs a fugue. In both cases, they produce variations upon themes. This chapter will examine common techniques for varying themes and some mathematics that helps describe how these techniques relate to each other. As we shall see, the mathematics not only serves a descriptive role, but it has also been deliberately exploited by some composers in creating their works.

Heightening the Scales

One of the simplest and most natural – and therefore commonest – ways to alter a bit of music is to shift it higher or lower in pitch. Organists at sports events use this device all the time. In order to stir up the crowd, an organist will repeat some musical snippet over and over, with each repetition higher than the previous one, usually by a step of one semitone. Figure 1 shows a couple of typical examples (CD Tracks 6 and 7).

a

b

Fig. 1

Those specimens exhibit fairly blatantly the technique of *transposing,* or shifting, a musical pattern. But transpositions may be deployed more subtly. Take the following melody from Edward Elgar's first *Pomp and Circumstance* march (CD Track 8), also known in Britain as the song "Land of Hope and Glory" and in the United States as "that piece they play interminably at every graduation ceremony" (see Figure 2).

Fig. 2

Elgar took measures 1–3 and lowered them by seven semitones (a fifth) to obtain measures 5–7. However, rather than deriving measure 8 from measure 4 in the same way, he simply carried over the note from measure 7. And then, instead of just continuing to transpose like the sports organist, Elgar altered the melody further (see Figure 3).

Fig. 3

These measures are still related to the original four, but not as closely as measures 5–8 are. In particular, measure 10 results from measure 2 by means of a five-semitone drop *and* a change of rhythm.

In brief, transpositions appear throughout music, from the opening measures of *Three Blind Mice* to those of Beethoven's *Symphony #5*. But at various times, these shifts may be applied more systematically or less so. As might be expected, the more systematic the usage, the more that mathematics comes into play.

As shorthand for "transposition by n semitones," musicologists often write "T_n". According to this notation, then, Figure 1 demonstrates the application of T_1, and Figure 2 that of T_{-7}. In the latter case, the negative number indicates a downward shift. T_0, of course, represents the action – or rather the inaction – of leaving the notes unaltered. To refer to the nonchange T_0 as a transposition, although it may seem dubious, turns out to be quite convenient.

What about T_{12}? This operation raises everything by an octave, thereby changing the pitch of every note but not altering any pitch classes. Consequently, how you treat T_{12} depends on your context – on whether you take into account pitch or just pitch class. In many cases (especially with so-called twelve-tone music, which we will soon consider), composers view transpositions in terms of pitch class. From this standpoint T_{12} doesn't change a thing. In other words, as far as pitch class is concerned, T_{12} is the same as T_0. And so are T_{24}, T_{36}, T_{-12}, and so forth.

Similarly, T_{13} and T_1 are the same in this pitch-class framework, since only an octave's worth of shifting distinguishes them from each other. In general, if m and n differ by a multiple of twelve, then T_m and T_n have the same effect on pitch classes, and henceforth I will consider them as the same transposition. In everyday life we treat clocks this way, too. Setting a basic wall clock thirteen hours ahead or even twenty-five hours ahead accomplishes nothing more than moving it one hour ahead. When two numbers m and n differ by a multiple of 12, mathematicians say that m and n are equal modulo 12 – or mod 12, for short (see Sidebar 1). Every whole number, mod 12, equals 0 or 1 or ... or 10 or 11. Just as with the hours on

In our dealings with transpositions, we will work modulo 12, but other situations call for other numbers. For example, a watch that keeps twenty-four-hour time follows a mod 24 system. Advancing the watch twenty-four hours changes nothing. However, a twelve-hour shift will turn noon to midnight and vice versa. In a different setting altogether, many car odometers register mileage mod 100,000. After a car so equipped racks up 100,000 miles, its odometer reading matches that of a brand-new vehicle.

Musically, it sometimes pays to work mod 7. Transpositions often consist of shifts by a certain number of notes of a scale, rather than a certain number of semitones. Since each major and minor scale contains seven different pitch classes, seven would fill the role that twelve will play for us.

a clock, then, we need only deal with the twelve transpositions T_0 through T_{11}.

Very often a transposed melody itself gets transposed. Consider again the "Charge!" example of Figure 1. The last pair of measures was obtained by performing two successive T_1 operations. But by skipping the middle pair of measures, we could have gone directly from the first pair to the last by using T_2. In general, following a T_m operation by a T_n yields the same net result as the single transposition T_{m+n}. Written more concisely, $T_m T_n = T_{m+n}$. Since we will work mod 12, this entails that, for example, $T_6 T_{10} = T_4$. The table of Figure 4 shows how all twelve transpositions combine with each other.

Through the Looking Glass

Both musically and mathematically, things become more interesting when transpositions combine with other kinds of operations as well. What else can a composer do to a musical fragment besides transpose

	T_0	T_1	T_2	T_3	T_4	T_5	T_6	T_7	T_8	T_9	T_{10}	T_{11}
T_0	T_0	T_1	T_2	T_3	T_4	T_5	T_6	T_7	T_8	T_9	T_{10}	T_{11}
T_1	T_1	T_2	T_3	T_4	T_5	T_6	T_7	T_8	T_9	T_{10}	T_{11}	T_0
T_2	T_2	T_3	T_4	T_5	T_6	T_7	T_8	T_9	T_{10}	T_{11}	T_0	T_1
T_3	T_3	T_4	T_5	T_6	T_7	T_8	T_9	T_{10}	T_{11}	T_0	T_1	T_2
T_4	T_4	T_5	T_6	T_7	T_8	T_9	T_{10}	T_{11}	T_0	T_1	T_2	T_3
T_5	T_5	T_6	T_7	T_8	T_9	T_{10}	T_{11}	T_0	T_1	T_2	T_3	T_4
T_6	T_6	T_7	T_8	T_9	T_{10}	T_{11}	T_0	T_1	T_2	T_3	T_4	T_5
T_7	T_7	T_8	T_9	T_{10}	T_{11}	T_0	T_1	T_2	T_3	T_4	T_5	T_6
T_8	T_8	T_9	T_{10}	T_{11}	T_0	T_1	T_2	T_3	T_4	T_5	T_6	T_7
T_9	T_9	T_{10}	T_{11}	T_0	T_1	T_2	T_3	T_4	T_5	T_6	T_7	T_8
T_{10}	T_{10}	T_{11}	T_0	T_1	T_2	T_3	T_4	T_5	T_6	T_7	T_8	T_9
T_{11}	T_{11}	T_0	T_1	T_2	T_3	T_4	T_5	T_6	T_7	T_8	T_9	T_{10}

Fig. 4

it? Play it backwards, for one thing. J. S. Bach's *A Musical Offering* (CD Track 9), for example, contains a piece based on this idea (Figure 5). The second half of the piece consists precisely of the first half played in reverse order, or *retrograde*. More precisely, the top (resp., bottom) line of the last nine measures is the retrograde of the bottom (resp., top) line of the first nine. (Incidentally, this selection was written out by Bach's pupil Johann Philipp Kirnberger, whom we will encounter again in Chapter 6 in the context of musical dice games.)

The last movement of Beethoven's *Piano Sonata #29 (Hammerklavier)* and music of Franz Joseph Haydn (see Sidebar 2) also feature large-scale use of such reversals. But in miniature the technique shows up in places like *Auld Lang Syne* (CD Track 10), in which – making allowances for rhythmic variations, including repeated notes – the FACD of "never bro't to mind" is mirrored by its retrograde DCAF in the following "Should auld acquaintance" (Figure 6).

Fig. 5

Should auld ac-quaint-ance be for-got, And nev-er bro't to mind? Should auld ac-quaint-ance

Fig. 6

SIDEBAR 2

Haydn's *Symphony #47* includes a minuet "al reverso." Haydn must have liked this minuet because he also used it in his *Sonata #4 for Piano and Violin* and in his *Piano Sonata #26.* So, not only did the retrograde give him two measures of music for every one that he wrote, but he also got extra mileage out of his efforts by recycling the minuet. (Admittedly, it takes more work in the first place to write a tune that sounds good both forwards and backwards!) For obtaining the most pieces from the fewest measures, however, this cannot begin to compete with the minuet trios from the dice game attributed to Haydn (Chapter 6).

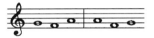

Fig. 7

Indeed, forming the retrograde of a melody corresponds, in written notation, to reflecting it left-to-right (see Figure 7). Notice, though, that there is a big perceptual difference between visual left–right reversal and auditory time reversal. People have no trouble recognizing objects in mirrors, but hardly anyone would observe, just from listening, the symmetry in the Bach example. Our brains tend not to do well at processing information, particularly information that we hear, in reverse chronological order. Try it: aloud read when sense little very makes sentence This! Compared to transposition, then, retrograde is a less noticeable operation and occurs less frequently.

On the other hand, people can often perceive the musical reflections called *inversions* fairly easily. To invert a melody, you flip it up–down, rather than left–right. In other words, for every step the original theme takes raising the pitch, the inversion will take a step of the same size, but lowering the pitch (and, of course, vice versa). Figure 8 illustrates an inversion of the same sequence of notes used in Figure 7. Here, the original steps – two semitones downwards,

Fig. 8

then four upwards – invert to steps of two upwards, then four down-wards.

Although less simple to describe than retrogrades, inversions present themselves more clearly to the ears, especially when the in-version and the original theme are played simultaneously. In that case the opposing motions of the two melodic lines contrast as strik-ingly in sound as they do in the printed notation. The opening of Sousa's march *The Thunderer* (CD Track 11) exhibits this contrast quite starkly. Notice that the two lines do not strictly follow the inver-sion pattern. For example, at the first step, the upper melody rises by two semitones, while the lower melody falls by only one. But the effect of the two lines steadily diverging from each other is unmistakable. Indeed, the approximate inversion relationship makes much more of an impression on our ears than the exact retrograde relationship that the pitches also possess (Figure 9).

Fig. 9

Brahms, in his *Variations on a Theme of Robert Schumann* (Figure 10 and CD Track 12), likewise juxtaposed a melody (top line) with an inversion of it (bottom line). Unlike Sousa, Brahms used an exact inversion here, but he made the effect more subtle by adding other notes in the intermediate range, as well as by using a melody that both rose and fell.

Fig. 10

Playing by the Rules

Having already analyzed how transpositions combine with each other, let's now toss retrogrades and inversions into the mix. We can obtain combinations without end, such as applying T_3 to a retrograde of T_5 applied to an inversion of the original theme. (For a conjunction of all sorts of these operations, see the excerpt from Liszt's *Hungarian Rhapsody #2* in Figure 11 and CD Track 13.) However, various such combinations duplicate each other in their net effect, just as $T_6 T_{10}$ and T_4 do. We will find, in fact, that putting transpositions, retrogrades, and inversions together yields a surprisingly small number of different results.

First, consider the retrograde operation, denoted R. One of its most basic traits is that it undoes itself – taking the retrograde of the retrograde of a theme leaves the theme unchanged. Thus, the operation RR accomplishes nothing; in symbolic shorthand, $RR = T_0$. Notice also that, for any transposition T_n, the operation RT_n equals $T_n R$. That is, if you list the notes in reverse order and then shift them, you get the same result as if you had first shifted and then reversed them (Figure 12). This unsurprising equation $RT_n = T_n R$ might induce a false sense of security. We shall soon see that inversions combine with transpositions according to *different* rules.

Fig. 11

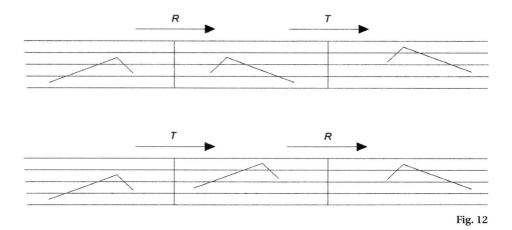

Fig. 12

Because of the above relationships, we can write any combination of R's and T's, no matter how complicated, as either a single transposition or a transposition followed by a retrograde. Take $RT_5 RT_2 RT_8$, for example. The $RT_n = T_nR$ rule lets us move all the T's to the left of the R's: $T_5 T_2 T_8 RRR$. By the $RR = T_0$ rule, this equals $T_5 T_2 T_8 T_0 R$, which is the same as $T_3 R$. Thus, only twenty-four different combinations of T's and R's can arise, the twelve possible T_n's and the twelve possible $T_n R$'s.

Inversions behave in a more complicated fashion than retrogrades. To begin with, a sequence of notes has only one retrograde, but several inversions. There is just one way to play something backwards: place a reflecting mirror at the end of it. But where do you place the mirror for an inversion? You have several choices. These correspond to different starting notes for the inverted version. Once you specify the note with which the inversion begins, the rest follow, but you are free to pick that initial note. For *The Thunderer* in Figure 9, Sousa began his pseudoinversion on the same pitch class as the theme, namely, C. In effect, Sousa placed his inverting mirror at C. Brahms in Figure 10, on the other hand, positioned his mirror differently; the theme started on F♯, the inversion on D. This locates the mirror at E, halfway between F♯ and D.

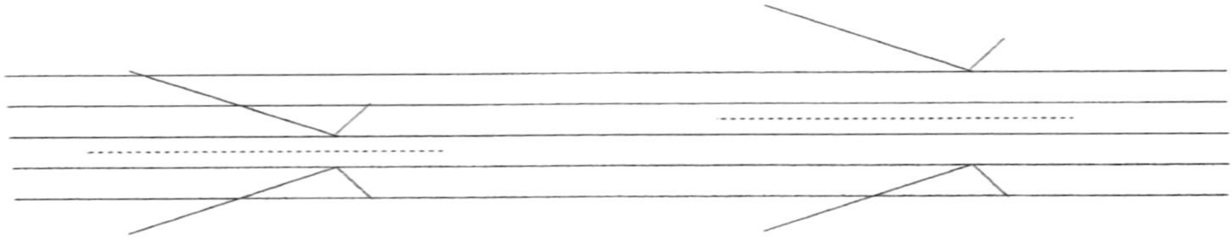

Fig. 13

As suggested by Figure 13, inversions in different mirrors will be transpositions of each other. So, since we already have transpositions at our disposal, we can select a single inversion and describe all inversions in terms of the designated one, along with transpositions. From now on we will use I to denote the inversion that results from a mirror placed at C. The effect of I, then, is to flip notes according to Figure 14. C♯ reflects to B and vice versa, and so on. Notice that not only does C remain unchanged by I, but so does F♯.

Now, we can consider the rules describing how I behaves in combinations of operations. Just as $RR = T_0$, likewise $II = T_0$ – again a double reflection in the same mirror doesn't change anything. And $IR = RI$, as illustrated in Figure 15. However, as I hinted earlier, IT_n will usually *not* be the same as T_nI. For example, take the "melody" consisting of the single note C and compare what happens to it under IT_2 and under T_2I. For IT_2 we first apply I to C, giving us C back again, and then apply T_2, yielding D. On the other hand, if we act on C with T_2I, we first use T_2, obtaining D, and then use I, which flips D to B♭. Figure 16 shows why IT_n will usually differ from T_nI.

Fig. 14

$$
\begin{array}{l}
\text{C} \leftrightarrow \text{C} \\
\text{C♯} \leftrightarrow \text{B} \\
\text{D} \leftrightarrow \text{B♭} \\
\text{D♯} \leftrightarrow \text{A} \\
\text{E} \leftrightarrow \text{A♭} \\
\text{F} \leftrightarrow \text{G} \\
\text{F♯} \leftrightarrow \text{G♭} \, (= \text{F♯})
\end{array}
$$

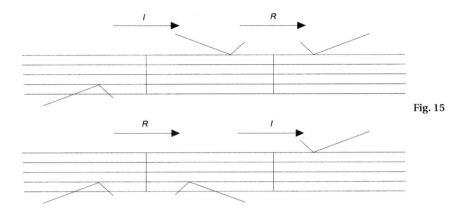

Fig. 15

All is not lost, however. The preceding example hints at the underlying pattern: although $IT_2 \neq T_2 I$, we have $IT_2 = T_{10}I$ and $T_2 I = IT_{10}$. In general, $IT_n = T_{12-n}I$. This fact can be verified with a little work by using the ideas behind Figures 13 and 16. Notice that $IT_6 = T_{12-6}I = T_6 I$ and that $IT_0 = T_{12-0}I = T_{12}I = T_0 I$. So, in these two cases IT_n does manage to equal $T_n I$.

By invoking the rules $IT_n = T_{12-n}I$ and $IR = RI$ along with the old rule $T_n R = RT_n$, we can rearrange any combination of T's, I's, and R's so as to lump all the T's together, all the I's together, and all the R's together. For instance, $T_7 IRT_4 I = T_7 IT_4 RI = T_7 T_8 IRI = T_7 T_8 IIR$.

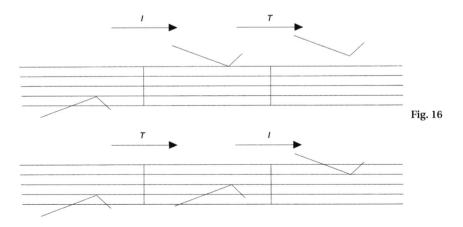

Fig. 16

THE MATH BEHIND THE MUSIC

With all the I's together, we can then remove pairs of I's because of the rule $II = T_0$, and similarly with the R's. Finally, we can combine the T's into a single T. Continuing the above example, $T_7 T_8 I I R = T_7 T_8 R = T_3 R$. In this manner, any combination whatsoever of T's, I's, and R's can be written as a T, possibly followed by I, possibly followed by R. That is, we can write the operation in one of these four forms: T_n, $T_n I$, $T_n R$, $T_n I R$. With twelve choices for n in each form, only forty-eight different operations exist. Any mix of T's, I's, and R's, no matter how complex, is one of these forty-eight. We could construct a table, similar to Figure 4, showing how all forty-eight operations combine with each other. Writing out the full 48-by-48 picture would certainly be very tedious, but we can easily fill in any particular slot in the table by using our rules; for example, $T_3 I$ combines with T_8 to yield $T_3 I T_8 = T_3 T_4 I = T_7 I$.

Musical Groups

Figure 4 is reminiscent of an addition table, and, indeed, the behavior of the forty-eight T-I-R operations parallels that of numbers in many respects.

Combining two operations results in an operation.	Adding two numbers results in a number.

We describe this situation by saying that numbers are *closed* with respect to addition. Likewise, the forty-eight operations are closed with respect to forming combinations.

For any operation X, both $X T_0$ and $T_0 X$ equal X.	For any number x, both $x + 0$ and $0 + x$ equal x.

An object that, like 0 or T_0, leaves everything unchanged is called an *identity element*. As another example, the number 1 serves as the identity element for multiplication of numbers.

Every operation X has an opposite operation Y with the property that XY and YX equal the identity T_0. For example, the opposite of T_5 is T_7, since $T_5 T_7 = T_0 = T_7 T_5$. The opposite of $T_2 I$ is $T_2 I$ itself, since $(T_2 I)(T_2 I) = T_0$.

Every number x has an opposite number y with the property that $x + y$ and $y + x$ equal the identity 0. For example, the opposite of 3 is -3, since $3 + (-3) = 0 = (-3) + 3$.

Roughly speaking, the opposite of something is what it takes to undo that thing. The usual mathematical word in this context is not "opposite," but "inverse"; however, to avoid confusion with musical inversions, I will stick with "opposite."

For any operations X, Y, and Z, we have $(XY)Z = X(YZ)$. For example, if $X = T_9 R$,

$\quad Y = T_5 I R$, $Z = T_3$,

$(XY)Z = [(T_9 R)(T_5 I R)] T_3$

$\quad = [T_2 I] T_3 = T_{11} I$,

and $X(YZ) = (T_9 R)[(T_5 I R) T_3]$

$\quad = T_9 R[T_2 I R] = T_{11} I$.

For any numbers x, y, and z, we have $(x + y) + z = x + (y + z)$. For example, if $x = 3$, $y = 2$,

$\quad z = 4$,

$(x + y) + z = (3 + 2) + 4 =$

$\quad 5 + 4 = 9$,

and $x + (y+z) = 3 + (2+4) =$

$\quad 3 + 6 = 9$.

Addition and T-I-R combination are said to be *associative*, since we can associate the middle term with either end term first and still get the same answer.

Many systems, both inside and outside mathematics, share the preceding properties with numbers and T-I-R operations. Abstracting from such situations, mathematicians call a *group* any collection of objects that is equipped with a means of combining those objects and that satisfies the following properties.

- The collection is closed with respect to combining its objects.
- The collection includes an identity element.
- Every object in the collection has an opposite in the collection.
- Combinations are associative.

SIDEBAR 3

One particularly important subgroup consists of the twenty-four T-I operations – the half of the forty-eight that do not use the retrograde operation. Known as a *dihedral* group, this subgroup may be interpreted geometrically in terms of the symmetry of a twelve-sided polygon. A related group will appear in Chapter 8 in conjunction with the configurations involved in contradances.

Fig. 17

T_0, R, I, IR

T_0, T_6

T_0, T_4, T_8, I, T_4I, T_8I

T_0 by itself

Notice that the collection of transpositions $T_0, \ldots, T_{11}$ forms a group in its own right. Combining two transpositions produces a transposition, so the collection is closed. It contains the identity T_0, and the opposite of any transposition is likewise a transposition and thus in the collection. Finally, since associativity holds for any of the forty-eight operations, it certainly holds for the transpositions. We call the transpositions a *subgroup* of the group of forty-eight operations. The key feature of a subgroup is that you can combine and take opposites of the subgroup's objects as much as you want, and you will always get a result that lies in the subgroup – you never escape into the other parts of the larger group. Figure 17 lists some of the many subgroups of the group of forty-eight operations (see also Sidebar 3).

Groups in Practice

Now this is all well and good, but has any composer ever really used groups and subgroups in writing music? In fact, several

Fig. 18

people have, especially those belonging to the so-called twelve-tone school. Twelve-tone compositional technique uses themes that are sequences of (not surprisingly!) twelve notes. But not just any such sequence. The notes must come from different pitch classes. Since there are twelve pitch classes, this means that each class occurs once and only once in the theme.

Some short twelve-tone pieces consist of a single such sequence (often called a *tone row*) played over and over. For example, the sequence EDABGB♭FCE♭D♭G♭A♭ forms the basis of Hanns Jelinek's *Two-Voice Invention on a Twelve-Tone Row* (Figure 18 and CD Track 14). Rhythmic alterations and the selection of different representatives from a pitch class help avoid monotony here. Lengthier compositions, however, call for the use of additional rows to provide sufficient variety. The classical twelve-tone strategy is to derive these extra rows by applying T-I-R operations to the original row. (Notice that any such operation will indeed turn a twelve-tone row into a twelve-tone row.) Thus, a typical piece based on a certain row can draw on any of forty-eight related rows, and the use of rows that are related induces some unity along with the variety.

Take, for instance, the opening of Arnold Schoenberg's violin concerto (Figure 19 and CD Track 15). The first few measures lay

Fig. 19

out the basic tone row AB♭D♯BEF♯CD♭GA♭DF. (As this passage illustrates, twelve-tone technique allows certain other modifications of the tone rows, such as playing some of the notes of the sequence simultaneously or repeating a portion of the row.) The following several measures produce a new row DC♯A♭CGFBB♭ED♯AF♯ by applying $T_1 I$ to the original. These two rows are then counterpointed against each other. Specifically, the violin (top staff) plays the original while the orchestra (bottom two staves) plays $T_1 I$. Then they swap, with the orchestra getting the original and the violin $T_1 I$. Next (five measures from the end of this excerpt) Schoenberg introduces the retrograde of the original row. Finally, with $T_1 I R$ of the original, he rounds things out in two senses. $T_1 I R$, of course, is the combination of the $T_1 I$ and the R previously used. But more than that, no further applications of $T_1 I$, R, or $T_1 I R$ will generate any new rows. In other words, the four rows that have appeared correspond to T_0, $T_1 I$, R, and $T_1 I R$ of the original row, and these four operations form a group – a subgroup of the group of forty-eight.

Not that every twelve-tone piece relies on groups and subgroups or even that every appearance of a group was specifically intended as such by the composer. One could reasonably interpret Schoenberg's use of $T_1 I R$ above as just a natural sequel to his use of $T_1 I$ and R, without having to invoke any mathematical theory of groups. However, some twelve-tone composers have quite explicitly organized works of theirs according to group-theoretic ideas.

One of the most prominent of such musicians is the Pulitzer Prize winner, Milton Babbitt. His writings, such as "Set Structure as a Compositional Determinant," expound the application of groups and related notions to the writing of music. Babbitt's pieces can involve groups in very refined and intricate ways.

For example, consider Figure 20 and CD Track 16, the start of the first of Babbitt's "Three Compositions for Piano." In each measure the bottom line contains six notes, as does the top line. The basic tone row is the bottom line of the first two measures: B♭E♭FDCD♭GBF♯AA♭E.

Fig. 20

Operating on this row with $T_3 I R$ generates the bottom half of measures 3–4. Likewise, applying $T_9 I$ and $T_6 R$ to the basic tone row produces the lower lines of measures 5–6 and 7–8, respectively. Just as happened with the Schoenberg example, T_0, $T_3 I R$, $T_9 I$, and $T_6 R$ form a subgroup. Meanwhile, the upper line also derives from the

The idea of coset is implicit in the familiar classification of whole numbers as even and odd. The even numbers constitute a subgroup of the group of numbers combined by addition, since the sum of two evens is even, and so on. If we form a coset by adding 1 to every even number, we get the collection of odd numbers.

Instead of the evens, that is, the multiples of 2, what if we take the subgroup consisting of the multiples of 12, such as 0, 12, 24, and -12? Then the coset obtained by adding 1 contains (among other numbers) 1, 13, 25, and -11, while the coset obtained by adding 5 includes 5 and -7. In fact, two numbers m and n will belong to the same coset exactly when $m = n \pmod{12}$. Cosets also underlie our earlier classification of the T-I-R operations into four types: T_n, T_nR, T_nI, and T_nIR. The twelve transpositions make a subgroup of the group of forty-eight, and combining these with R, for example, yields the twelve T_nR's as a coset.

basic tone row in two-measure chunks, via T_6, R, T_9IR, and T_3I, respectively. There are two things to note about these four additional operations. First, although they do not constitute a subgroup themselves, they are the results of combining the original subgroup of four operations with T_6: $T_6 = T_6T_0$, $R = T_6(T_6R)$, $T_9IR = T_6(T_3IR)$, and $T_3I = T_6(T_9I)$. In mathematical jargon, T_6, R, T_9IR, T_3I form a *coset* of the subgroup (see Sidebar 4). Second, the eight operations in the subgroup and coset taken together constitute yet another subgroup.

Observe also that Babbitt chose the tone row and the operations so that within each single measure the twelve notes of the two lines yield a tone row. And, in fact, he smuggled further group-theoretic relationships within the basic tone row. On top of all this going on with the pitch classes, Babbitt echoed the operations in his rhythmic structures, too. For example, the basic tone row appears in measures 1–2 broken up rhythmically into clumps of 5, 1, 4, and 2 notes. When the retrograde (or a transposed retrograde) of the row occurs, the rhythmic pattern also reverses: 2, 4, 1, 5. When an inversion occurs, the pattern is inverted by reflecting the numbers in a mirror placed at 3, turning 5, 1, 4, 2 into 1, 5, 2, 4. Naturally, when the basic tone row is both inverted and reversed, Babbitt used a 4, 2, 5, 1 rhythm.

And all this just in the first eight measures! Certainly, among composers Babbitt's interest and background in mathematics is exceptional – he even taught university math. However, he has not been alone in relating groups to music, and several passages in standard musicological books on twelve-tone composition read like math texts.

But then group theory has many applications in various disciplines. Any time you analyze how things, particularly operations, combine with each other, groups probably lurk in the background. So, it's really not too extraordinary that groups crop up in music. In fact, the next chapter will show how, in a totally different context, musicians developed many group-theoretic ideas, not only long before the twelve-tone school, but even before mathematicians themselves abstracted the notion of a group.

FIVE

BELLS AND GROUPS

At first, the style of music known as change-ringing might not seem to have much going for it. A composition for piano can draw on eighty-eight different notes. An orchestral piece has even more notes available, plus the wide range of sonorities to be found in the various instruments. In contrast, change-ringing might use only six or seven notes, all played on tower bells. But within such an apparently restrictive framework, ringing the changes has developed into a highly refined art form. Its centuries-old tradition continues today in thousands of bell towers, especially in England.

Counting the Change

How is change-ringing music organized to yield something interesting out of such sparse material? Just as with the twelve-tone music described in the last chapter, change-ringing relies on sequences consisting of each available note played exactly once. But there are significant differences. First, instead of the twelve pitch classes of the chromatic scale, change-ringing uses a smaller set of notes, as mentioned earlier. In fact, the traditional notation for such music does not even specify the pitches to be used. The notes are simply designated $1, 2, 3, \ldots$, where 1 indicates the highest-pitched bell, 2 the second highest, and so on.

Moreover, change-ringing typically involves not just a comparative handful of sequences, but rather *all* possible sequences of the notes. Consider, for example, three bells. There are six different arrangements we can form: 123, 132, 213, 231, 312, and 321. Each such arrangement is called a *change*, giving rise, of course, to the name "change-ringing." With these changes as building blocks, we can then put together the following piece: 123213231321312132123. The structure of this piece becomes more transparent when the changes are written one to a line.

123

213

231

321

312

132

123

Notice that the first and last changes are the same, consisting of the bells played in order from highest to lowest. In between those, all other changes occur, each exactly once.

In general, a so-called *extent* on a given number of bells comprises the various changes on those bells, organized according to certain rules. An extent on n bells begins and ends with the change $12 \ldots n$. And in between, each other n-bell change will appear once and only once.

A question arises naturally from this: with n bells, how many changes are there? The answer will determine how long it takes to perform a complete extent. Clearly, with one bell only one change exists, namely, 1. We can form the two changes 12 and 21 out of two bells. And as we've already seen, three bells can be arranged into six different changes.

n	$1 \times \cdots \times n$
1	1
2	2
3	6
4	24
5	120
6	720
7	5040
8	40,320

Fig. 1

Notice that any change on bells 1, 2, 3 must contain within it the bells 1 and 2 in some order. If that order is 12, then we can add bell 3 at the beginning, middle, or end to obtain 312, 132, or 123. Likewise, if 1 and 2 appear in the order 21, then again bell 3 can go in one of three slots to produce 321, 231, or 213.

This observation points the way to the general pattern. The changes on bells 1, 2, 3, 4 arise by expanding each of the three-bell changes in four different ways. For example, adding bell 4 to the change 312 yields 4312, 3412, 3142, or 3124. Thus, each of the 6 three-bell changes leads to 4 four-bell changes, for a total of $6 \times 4 = 24$. Similarly, each four-bell change gives rise to 5 five-bell changes, so there are $24 \times 5 = 120$ ways to arrange five bells. In general, the number of changes on n bells is $1 \times 2 \times \cdots \times n$, the product of numbers from 1 through n.

The table in Figure 1 shows that as the number of bells increases, the number of changes grows dramatically. For example, Track 17 on the CD contains a brief excerpt of the seven-bell extent known as Grandsire Triples. A seven-bell extent comprises 5,040 different changes plus a repeat of the change 1234567. With seven bells per change, the extent calls for $5,041 \times 7 = 35,287$ rings. At a rate of roughly five rings per second, this takes about two hours

Fig. 2a

to perform. A similar calculation yields a time of roughly eighteen hours for an extent on eight bells. Needless to say, eight-bell extents are not performed very often. (By the same reasoning, there are $1 \times 2 \times \cdots \times 12 = 479,001,600$ twelve-tone rows. Twelve-tone composers need not fear running out of material.)

Travel Restrictions

Another rule for ringing changes stems from the physical limitations of playing bells. A tower bell can weigh as much as four tons (Figure 2). Consequently, each individual ring of a bell takes a certain amount of time to accomplish, around two to three seconds. Objects that heavy just don't move very quickly! (See Sidebar 1.) So, you cannot repeat a note on a tower bell as fast as you can on more everyday instruments. For that reason it wouldn't do to follow the change 314652 immediately with the change 214653. The necessary time lag between the back-to-back occurrences of bell 2 would interrupt the steady flow of notes. In fact, the change following

Fig. 2b

314652 should really have bell 2 either in the last or the next-to-last position to allow enough time.

Due to such considerations, ringers follow this adjacency rule: In going from one change to the next, each bell either remains in the same position or else swaps positions with a bell that was adjacent to it – either immediately before it or immediately after it.

59

S I D E B A R   1

A change-ringing performance involves several people: one for each bell, plus a conductor. The length of time it takes to ring a bell – accomplished by pulling on a rope – means that a player must act well before the note will sound. Under these circumstances, how do people know when to pull their ropes? By watching the motions of the other ropes. The skill of determining when to play by this visual clue, without getting confused by the out-of-synch auditory feedback, is known to ringers as *ropesight*. Along with the direction of the conductor, ropesight enables ringers to play this complex music properly.

For example, 314652 can be followed by 134562, in which the adjacent 3 and 1 swap places, as well as the adjacent 6 and 5. Likewise, 314652 could be followed by 341652 or 136425, but not by 413652 or 214653. Figure 3 illustrates these five cases in respective order. The adjacency rule thus ensures that in a six-bell extent, any two successive rings of a given bell are separated by at least four rings of other bells, thereby giving the ringer enough time to set the bell in motion.

The three-bell extent given at the beginning of this chapter obeys the adjacency rule. Even so, in that extent successive rings of any particular bell come close together. As a result, the extent on three

Allowed　　　　　　　　**Not allowed**

Fig. 3

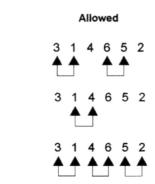

bells must be performed at a comparatively slower pace. But then ringers can afford to play such an extent slowly. The adjacency rule permits long extents to proceed at full speed.

As a puzzle at this point, you might want to try to design a four-bell extent that conforms to the adjacency rule. Such extents exist, although it is not obvious in advance that they do. In fact, it isn't even obvious that the restriction to swaps of adjacent bells allows you to obtain all possible changes, much less get each one exactly once before ending up on the change with which you began.

But you can. No matter how many bells, you can convert any change into any other just by swapping neighbors. A foolproof strategy is to put bells into their desired positions, one at a time, starting with the rightmost position. Say, for instance, that you want to turn 123456 into 532614. First, move 4 to the rightmost slot by swapping it with its right-hand neighbor twice.

$$1\ 2\ 3\ 4\ 5\ 6$$
$$\updownarrow$$
$$1\ 2\ 3\ 5\ 4\ 6$$
$$\updownarrow$$
$$1\ 2\ 3\ 5\ 6\ 4$$

Next, move 1 rightwards, again swap by swap.

$$1\ 2\ 3\ 5\ 6\ 4$$
$$\updownarrow$$
$$2\ 1\ 3\ 5\ 6\ 4$$
$$\updownarrow$$
$$2\ 3\ 1\ 5\ 6\ 4$$
$$\updownarrow$$
$$2\ 3\ 5\ 1\ 6\ 4$$
$$\updownarrow$$
$$2\ 3\ 5\ 6\ 1\ 4$$

As a bonus, those operations also put 6 in the right place, so now move 2 two notches and, finally, 3.

2 3 5 6 1 4

$\updownarrow$

3 2 5 6 1 4

$\updownarrow$

3 5 2 6 1 4

$\updownarrow$

5 3 2 6 1 4

Bell Letters

Before going any further, we need a less cumbersome method of describing the action of a swap on a change. On three occasions in the preceding example, we swapped the two leftmost bells with each other. It will prove convenient to have a shorthand way to refer to such swaps of certain positions, regardless of which bells happen to inhabit those positions at the moment. Label the positions A, B, C, . . . from left to right. (AB) will then denote the swap of the bells in positions A and B, that is, the two leftmost. Our example used (AB) when slots A and B were filled by bells 1 and 2, by bells 2 and 3, and by bells 3 and 5. Likewise, (BC) indicates the swap of the bells that lie second and third from the left, and so on. The example began with (DE) followed by (EF). In full, we performed the sequence of swaps (DE)(EF)(AB)(BC)(CD)(DE)(AB)(BC)(AB).

Notice the parallels with the transpositions, inversions, and retrogrades of the last chapter. Just as T-I-R operations act on twelve-tone rows, so do swaps act on changes. And just as we can combine T-I-R operations by applying them successively, so too can we combine swaps into sequences like the one above. Most importantly, the sequences of swaps form a group. Group theory, in fact, permeates

Composers like Milton Babbitt have used facets of the well-developed mathematical theory of groups in their compositions. On the other hand, change-ringing came on the scene long before group theory did. Fabian Stedman (1640–1713) systematized the art of ringing changes, finding much structure in the patterns. This chapter touches on only a fraction of that structure, which can now be most conveniently described in terms of groups. However, mathematicians did not formulate the general concept of a group until the 1800s. Stedman's findings, which anticipated the abstract theory by well over a century, were expressed in the empirical context of bells.

SIDEBAR 2

the design of change-ringing extents. (From a historical viewpoint, however, bells present a different situation than twelve-tone music does; see Sidebar 2.)

Recall that every group must contain an identity element, that is, an object that leaves unaltered whatever it combines with. We considered the "do-nothing" operation in the T-I-R context as a transposition by zero semitones. Similarly, for bells we can describe the identity as a sequence of zero swaps. If this seems dubious, we can alternatively write the identity as (AB)(AB), since that double swap has no net effect. More generally, (BC)(BC) or (CD)(CD) or any other swap combined with itself produces the identity. This means that every swap is its own inverse. (Remember that two operations are inverse, or opposite, if their combination yields the identity.) You can easily verify that the inverse of any sequence of swaps is simply the sequence written in reverse order. For instance, the inverse of (BC)(AB)(BC)(CD) is (CD)(BC)(AB)(BC), since they combine to form

$$\text{(BC)(AB)(BC)}\underline{\text{(CD)(CD)}}\text{(BC)(AB)(BC)}$$
$$= \text{(BC)(AB)}\underline{\text{(BC)(BC)}}\text{(AB)(BC)}$$

$$= (BC)\underline{(AB)(AB)}(BC)$$
$$= (BC)(BC)$$
$$= \text{identity}$$

Patterns and Cosets

As we have seen, any rearrangement of bells can be accomplished by a suitable sequence of swaps. So unlike the $T\text{-}I\text{-}R$ group, which can generate only a small fraction of the twelve-tone rows from any given row, the swap sequences generate all possible changes. Consider again our extent on three bells. This extent proceeds from one change to the next via either (AB) or (BC).

$$
\begin{array}{ccc}
1 & 2 & 3 \\
(AB) & & \\
2 & 1 & 3 \\
& (BC) & \\
2 & 3 & 1 \\
(AB) & & \\
3 & 2 & 1 \\
& (BC) & \\
3 & 1 & 2 \\
(AB) & & \\
1 & 3 & 2 \\
& (BC) & \\
1 & 2 & 3 \\
\end{array}
$$

Thus (AB) converts 123 to 213, (AB)(BC) converts 123 to 231, (AB)(BC)(AB) converts 123 to 321, and so forth. The sequence (AB)(BC)(AB)(BC)(AB)(BC) – $((AB)(BC))^3$, for short – is the identity, since it effects no net alteration. Summarizing, the sequences (AB), (AB)(BC), (AB)(BC)(AB), (AB)(BC)(AB)(BC), (AB)(BC)(AB)(BC)(AB), and identity constitute the group of all possible ways to rearrange the bells. In mathematical jargon, rearrangements are called *permutations*, and so these six sequences are said to form the full permutation group on three bells.

The situation plays out somewhat differently with four bells. If we try alternating (AB)(CD) with (BC), for instance, we get the following.

```
        1   2   3   4
       (AB)     (CD)
        2   1   4   3
           (BC)
        2   4   1   3
       (AB)     (CD)
        4   2   3   1
           (BC)
        4   3   2   1
       (AB)     (CD)
        3   4   1   2
           (BC)
        3   1   4   2
       (AB)     (CD)
        1   3   2   4
```

At this point, another (BC) would reinstate the starting change, after having gone through only eight changes, rather than all twenty-four. In group-theoretic terms, $((AB)(CD)(BC))^4$ is the identity, and the various combinations of (AB)(CD) and (BC) make up a subgroup with only eight members. However, we can escape from that subgroup by throwing in a single (CD) swap. This moves us out of the subgroup and into one of its cosets (see Sidebar 3). Now resuming

S I D E B A R   3

Recall from the previous chapter that we obtain a coset of a subgroup by taking some designated member of the group and combining it with all the various members of the subgroup. But in general, the order in which we combine things makes a difference. $(AB)(BC) \neq (BC)(AB)$, just as $T_2I \neq IT_2$. So strictly speaking, there are two kinds of cosets, called *left* and *right* cosets. A left (resp., right) coset consists of combinations in which the designated group member is written on the left (resp., right) of the subgroup member. Usually, the left and right cosets differ from each other. I've been using left cosets and, for simplicity, will continue to refer to them simply as cosets.

the pattern of alternating (AB)(CD) and (BC), we obtain the other members of that coset.

$$
\begin{array}{cccc}
1 & 3 & 2 & 4 \\
& & \multicolumn{2}{c}{(CD)} \\
1 & 3 & 4 & 2 \\
(AB) & & (CD) & \\
3 & 1 & 2 & 4 \\
& (BC) & & \\
3 & 2 & 1 & 4 \\
(AB) & & (CD) & \\
2 & 3 & 4 & 1 \\
& (BC) & & \\
2 & 4 & 3 & 1 \\
(AB) & & (CD) & \\
4 & 2 & 1 & 3 \\
& (BC) & & \\
4 & 1 & 2 & 3 \\
(AB) & & (CD) & \\
1 & 4 & 3 & 2 \\
\end{array}
$$

Again now, (BC) would lead to a repeat of an earlier change, but (CD) will push us over into yet another coset, and we can continue.

$$
\begin{array}{cccc}
1 & 4 & 3 & 2 \\
& & \text{(CD)} & \\
1 & 4 & 2 & 3 \\
\text{(AB)} & & \text{(CD)} & \\
4 & 1 & 3 & 2 \\
& \text{(BC)} & & \\
4 & 3 & 1 & 2 \\
\text{(AB)} & & \text{(CD)} & \\
3 & 4 & 2 & 1 \\
& \text{(BC)} & & \\
3 & 2 & 4 & 1 \\
\text{(AB)} & & \text{(CD)} & \\
2 & 3 & 1 & 4 \\
& \text{(BC)} & & \\
2 & 1 & 3 & 4 \\
\text{(AB)} & & \text{(CD)} & \\
1 & 2 & 4 & 3 \\
& & \text{(CD)} & \\
1 & 2 & 3 & 4 \\
\end{array}
$$

The final (CD) brings us home to 1234 after having visited all the other changes.

The cosets of subgroups turn out to be quite handy in designing extents. They figure prominently in the structures of many extents, especially those on larger numbers of bells. Their usefulness stems in large part from the fact that the cosets of any subgroup of a group (counting the subgroup itself as one of its cosets) break up the whole group into several nonoverlapping pieces. Figure 4 schematically depicts the general situation. In our case with four bells, the twenty-four-member permutation group decomposed into a subgroup and its two cosets. Furthermore, each of those pieces has eight members.

Fig. 4

Subgroup	Coset	Coset	Coset	Coset	Coset
Coset	Coset	Coset	Coset	Coset	Coset

This equitable division holds in general – each coset of a subgroup contains the same number of objects as the subgroup. (With the appropriate interpretation of "number," the preceding statement remains true even if the group has infinitely many members.) As a by-product, it follows that the number of objects in the whole group must be a multiple of the number of objects in a subgroup. Thus, neither the group of permutations of four bells (24 members) nor the T-I-R group (48 members) can include a subgroup consisting of exactly five objects.

Evening Bells, Odding Bells

Other group-theoretic notions also come into play in the construction of extents. One very important idea ties together different ways of expressing the same rearrangement. We have already seen several descriptions of the identity: $(AB)(AB)$, $((AB)(BC))^3$, $((AB)(CD)(BC))^4$, as well as the sequence of zero swaps. Notice the number of swaps in each of these different representations of the identity: 2, 6, 12, and 0. These are all even numbers. Rather than being a coincidence, this is part of a general pattern. Any such expression for the identity must consist of an even number of swaps. No matter how hard you try, you can't write the identity as a combination of, say, 999 swaps. On the other hand, you can write certain other permutations only with an odd number of swaps. For instance, $(AB)(BC)(AB)$ can also be written as $(BC)(AB)(BC)$ or as $(DE)(AB)(DE)(BC)(DE)(AB)(DE)$, but never as a combination of 1,000 swaps. Every permutation, as a matter of fact, can be classified in this way as either even or odd. You can express the even (resp., odd) permutations only in terms of an even (resp., odd) number of swaps.

Furthermore, exactly half the permutations on a given set of bells are even and half are odd. The even permutations form a subgroup. (Combining an even number of swaps with an even number of swaps results in an even number. And the inverse of an even sequence of swaps is even, since you can form the inverse by just reversing the original sequence.) Combining all the even permutations with a single swap yields the odd permutations as a coset.

These even–odd facts entail various consequences for the design of extents. For one, if you always go from change to change by means of an even number of swaps, you won't get a full extent. So, for example, you cannot build a five-bell extent on just the moves (AB)(CD), (AB)(DE), and (BC)(DE). For another consequence, consider an extent that proceeds one swap at a time. If each change is derived from the previous one by a single swap, then the extent will alternate even and odd arrangements. Thus, starting from 12345, the change 21345 might conceivably appear first after 12345, or third after it, and so on, but never second or fourth or 100th after it.

Several other aspects of group theory show up in change-ringing. Although not a creation of professional mathematicians, the art of ringing changes relies on quite sophisticated mathematical underpinnings. In lieu of traditional melodies, this music derives much of its beauty from the richness of structure hidden within such simple material as the arrangements of a small number of notes.

SIX

MUSIC BY CHANCE

Although most music is composed with more or less deliberateness, exceptions arise. Consider wind chimes. Of course, the maker of a set of wind chimes determines the pitches of the notes that are sounded. But as to which note plays when – that varies unpredictably with the breeze. The music is *random*.

A less common instrument, the Aeolian harp, likewise operates in a random, wind-driven fashion. However, randomness in music can occur in other forms, as well. Various composers have used chance events such as dice rolls to help produce music that is performed by humans on standard instruments. Such randomized compositions flourished in the eighteenth and twentieth centuries, in particular. In this chapter, we will sample some of these experiments in chance-controlled music.

Music on a Roll

In the second half of the eighteenth century, there arose a virtual epidemic of musical dice-rolling games, sort of the musical equivalent of paint-by-number kits. Johann Philipp Kirnberger set this fad in motion in 1757 with his very first published work. Other composers quickly followed suit. Over the next fifty years, many musical dice games appeared, some of them marketed as being penned by no less figures than C. P. E. Bach, Haydn, and Mozart.

Kirnberger's publication bore the comparatively modest title of "Der allezeit fertige Menuetten- und Polonoisen-komponist" ("The ever-ready composer of minuets and polonaises"). The name of Piere Hoegi's musical dice game was more characteristic of the genre: "A Tabular System Whereby the Art of Composing Minuets Is made so Easy that Any Person, without the least knowledge of Musick, may compose ten thousand, all different, and in the most Pleasing and Correct Manner." Despite its verbosity, Hoegi's title underestimates his case. We will see that a typical game yields pieces numbering in the trillions. On the other hand, the Haydn game's claim of "un infinito numero di minuettie trio" ("an infinite number of minuet trios") goes a little too far!

How do these games work? In brief, the composer has written several bits of music, and you, as the player, roll dice to see which of the bits you play. Let's look at two of the games in more detail. Kirnberger's game produces pieces of dance music called polonaises. Each polonaise produced consists of fourteen measures, played one after the other. Kirnberger composed eleven different versions of each measure. In other words, you can use any of eleven alternatives for the first measure, then any of another eleven alternatives for the second measure, and so on. For each measure, the eleven options all employ the same harmonic scheme. In this way Kirnberger ensured that, no matter what selections are made, everything fits together into a pleasant-sounding polonaise. And how do you select the measures? That's where the dice come in. Roll a pair of dice, and the total will be a number from 2 through 12. Kirnberger's game includes a chart that specifies, for each measure, an alternative corresponding to each total of the dice.

Around 1790, a similar but simpler game was marketed as a composition of Franz Joseph Haydn, although whether Haydn actually wrote the music is somewhat in doubt. In this game you produce minuet trios (see Sidebar 1) by rolling a single die. Thus, you choose among six alternatives for each measure, the trio consisting of

Why is a piece for solo piano called a trio? Music for the courtly dance called the minuet usually comes in ABA form. In other words, the first ("A") melody is followed by a second ("B") melody, and then the A melody is repeated. Traditionally, the B melody appeared in a different key than the A melody and was written in three-part harmony, as if for a trio of performers. This middle section then became known as the trio, and the name stuck. And thus even though the piece by Haydn does not feature three-part harmony, it is still called a trio. In fact, the name carried over from minuets to pieces such as marches and piano rags. Such pieces usually come in forms more complex than ABA, but they typically shift into a different key about halfway through. The section in that new key is also referred to as a trio.

sixteen measures in all. Figure 1 displays the music in a 6-by-16 grid. In the first column are the six possible choices for the first measure, in the second column the six alternatives for the second measure, and so on. So, for example, you can play a minuet trio by reading straight across the top row. This would correspond to rolling sixteen 1's in a row with the die. That particular trio appears on Track 18 of the CD. Track 19 contains what you get if you roll sixteen 2's, while Track 20 plays the results of rolling 1, 2, 1, 2, 1, 2, and so forth.

The original presentation of the trio game camouflaged this simple structure a bit. Instead of organizing the music into a 6-by-16 grid with all of the alternatives for each measure lying in the same column, the publishers strung out the ninety-six measures in a row. And they jumbled the order of the measures. (The small number printed inside each measure in the grid corresponds to that measure's place in the original strung-out row.) A chart then guides you through the long row. For example, the chart says that for the first measure, if you roll 1, 2, 3, 4, 5, 6, you select, respectively, measure #72, 56, 75, 40, 83, 18. All of this, of course, is an unnecessarily complicated way of

To perform: start from the left and select any one bar from each succeeding column. To compose with dice, use the lower table on the front cover. Trios to dice-music minutes.

Fig. 1a

A complete continuous performance of these trios, at a steady minuet tempo, would take just under 900,000 years (without repeats) attributed to Joseph Haydn.

Fig. 1b

doing business – and serves the sole purpose of making the whole process seem more mysterious and magical.

Trillions of Trios

What makes the musical dice games amazing is how many pieces you can get from so few measures. Take Haydn's trios, for instance. You can choose the first measure in six different ways and then, for each of those choices, go on to select the second measure in any of six different ways. Thus, each of the six first choices spawns six further ways to go, resulting in six sixes or $6 \times 6 = 36$ ways to select the first two measures.

Often, situations such as this are depicted by a branching diagram (Figure 2). The diagram clearly indicates the thirty-six ways to put together the two successive choices. Now each of these thirty-six ways will branch out six further ways at the third measure. Thus, you can select the first three measures in $36 \times 6 = 216$ different ways, and so on.

In general, say that you have a sequence of choices to make. If you know how many options you have for each choice, you can compute how many ways you have of putting the whole packet of choices together. Just multiply the respective numbers of options. This rule, the so-called *Multiplication Principle*, is very useful in a variety of counting situations. In fact, we already used it in Chapter 5. The formula that calculates the number of arrangements of a collection of notes is just a particular case of the Multiplication Principle. Namely, in arranging n notes, you make a sequence of choices – which note goes first, which of the remaining notes goes second, etc. There are n ways to make the first selection, $n-1$ for the second, and so on. This implies that there are $n \times (n-1) \times \cdots \times 1$ different arrangements, as we have seen.

In the case of the Haydn trios, the sixteen measures give rise to sixteen factors of six to be multiplied together, so the total number

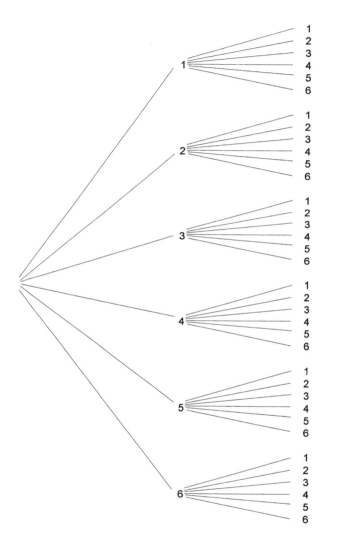

Fig. 2

of possible trios is

$$6 \times 6 \times \cdots \times 6 = 6^{16} = 2{,}821{,}109{,}907{,}456.$$

Quite a few trios for only ninety-six measures! Actually, I've lied just
a little. If you look closely at the music, you can see that there are not

77

six different options for the last measure. Haydn got a little lazy and only wrote three different versions, using one of those versions for four different rolls of the die. Similarly, you have just four different choices available for the eighth measure. So, the number of trios is really "only"

$$6 \times 6 \times 6 \times 6 \times 6 \times 6 \times 6 \times 4 \times 6 \times 6 \times 6 \times 6 \times 6 \times 6 \times 6 \times 3,$$

or 940,369,969,152. If you roll a die and produce a trio, you will almost certainly obtain one that nobody has ever picked before – nor is anyone likely to duplicate it exactly in the future. Kirnberger's game allows for an even greater number of polonaises: 11^{14}, which works out to nearly 380 trillion.

There is an important difference between using one or two dice in the musical games, although this difference may not be obvious at first glance. Consider Haydn's trios (one die) and Kirnberger's polonaises (two dice). For simplicity, pretend that Haydn didn't get lazy and wrote six truly different alternatives for each measure. Then *each of the 6^{16} possible trios is just as likely to occur as any other, but the 11^{14} polonaises are* not *equally likely.*

Let's look at that fact in more detail. On any single roll of one die, each of the numbers from 1 through 6 has as good a chance of coming up as any other number. So in the long run, we expect 1 to come up 1/6 of the time, 2 to come up 1/6 of the time, and so on. If you roll a die six million times, you expect to get around a million of each number. The usual jargon is to say that each number occurs with probability 1/6.

Because on each roll of a die any number from 1 through 6 is equally likely, it follows that any sequence of sixteen numbers is as likely to result from sixteen rolls as any other such sequence. In our context, any one of the 6^{16} Haydn trios stands just as good a chance of turning up as any other one – 1 in 6^{16}. In other words, the probability of selecting any particular trio is $1/6^{16}$.

Probability Out on a Limb

We can also arrive at that probability figure by another line of reasoning, which will prove useful later in other situations. Say we want to find the probability of rolling some particular sequence of numbers like 4, 2, 5, ... One sixth of the time your first roll will be a 4. One sixth of that 1/6 of the time, the second number will be 2. In other words, $1/6 \times 1/6 = 1/36$ of the time, the sequence starts 4, 2. Now one sixth of that 1/36 of the time, that is, $1/6 \times 1/6 \times 1/6$ of the time, the third number will be 5, and so on. You can expect to get your particular sequence of sixteen numbers $1/6 \times 1/6 \times \cdots \times 1/6$ of the time, with sixteen factors of 1/6 here. Thus, the probability is $(1/6)^{16} = 1/6^{16}$, as we computed before.

You can see that a Multiplication Principle for probabilities comes into play here, although I was phrasing it in terms of such-and-such a fraction of the time. This principle shows up most clearly in terms of a branching picture like the one we used before. Everywhere that branches lead to different alternatives, label each branch with the probability that its corresponding alternative is selected. The Probability Multiplication Principle says that to find the probability of reaching any point on the "tree," you multiply the probabilities on the branches leading to that point. In our case the picture looks like Figure 3, since for each die toss, every number has probability 1/6. The probability of rolling 4, then 2, for the first two rolls is $1/6 \times 1/6$, and so on.

Branching diagrams and multiplication of probabilities likewise explain how the two-die situation works. As every gambler knows, when you roll a pair of dice, you are much more likely to get a total of 8 than a total of 2. To help understand what happens, it is traditional to suppose that the two dice are colored differently, say one red and one green. The red die comes up as any number from 1 through 6 equally likely, and the same holds for the green die. Now, we may use the branching picture of Figure 3. In this context, we think of the

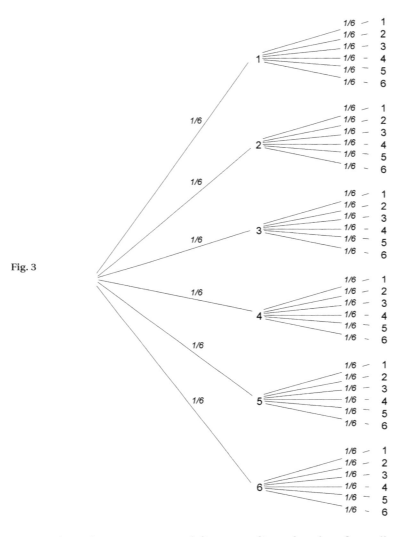

Fig. 3

picture's two stages as red die, green die, rather than first roll, second roll. But the options and their probabilities remain the same. So, for example, the probability that red comes up 1 and green comes up 1 is $1/6 \times 1/6 = 1/36$.

Well, red 1–green 1 is the only way that you can roll a total of 2, so the probability of totalling 2 is $1/36$. On the other hand, you can total 8 in five different ways: red 2–green 6, red 3–green 5, red 4–green 4, red 5–green 3, and red 6–green 2. Each such possibility, say red 3–green 5, has probability $1/6 \times 1/6 = 1/36$, as the branching diagram indicates, and thus occurs $1/36$ of the time. With five such possibilities, a total of 8 is achieved five thirty-sixths of the time, that is, the probability of the pair of dice coming up 8 is $5/36$.

Notice that here we added the five $1/36$'s together, rather than multiplying them. Those five $1/36$'s did not represent successive steps along one branch path, but rather different paths, any one of which yielded the total of 8. You can similarly compute the probabilities of the other totals, and these probabilities, in turn, may serve to label branches of a tree for Kirnberger's polonaise game (Figure 4). The probability of rolling fourteen 12's in a row is $\frac{1}{36} \times \frac{1}{36} \times \cdots = (\frac{1}{36})^{14}$. The probability of rolling fourteen 7's in a row is $\frac{6}{36} \times \frac{6}{36} \times \cdots = (\frac{6}{36})^{14}$. In other words, the most likely polonaise occurs 6^{14} times as often as the least likely, a factor of over 78 billion. Notwithstanding that, $(6/36)^{14}$ is about 0.000000000013, so even the likeliest polonaise is still a rare occurrence in a sea of possibilities.

The Twentieth Century

The heyday of the dice games coincided with a more general interest in music with a mathematical flavor. (Leonard Ratner's article, cited in the bibliography, discusses some other manifestations of that trend, from the construction of musical themes to the large-scale organization of compositions.) But by the early 1800s, the craze for these games faded out, and so did the use of randomness in Western art music. The Romantic Period flowered in the nineteenth century. Great stock was placed in the idea of music as conveying the composer's emotions and innermost self. Such works as Berlioz' *Symphonie Fantastique* or Richard Strauss' tone poems celebrate music as not just self-expression, but even autobiography. This subjective conception of music, naturally, predisposes composers not

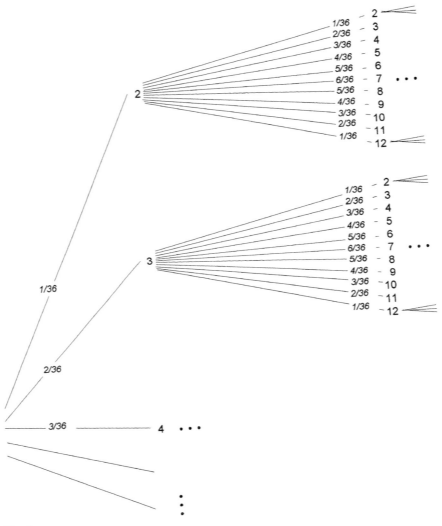

Fig. 4

to cede control of the music to external factors such as the roll of a die. Chance-driven music languished in the 1800s.

By the mid-twentieth century, however, this Romantic view exerted less influence on composers. And the time for random music was once again ripe for several other reasons, as well. One factor was a reaction against serial music. As mentioned in Chapter 4, composers extended Schoenberg's ideas to aspects of music beyond pitch. Thus, they meticulously organized pitch, rhythm, dynamics, timbre, and so on according to very strict patterns. Many found this too rigid and confining. To relinquish all that regimentation and turn over the reins to chance seemed like a liberation to some composers.

The ceding of control by the composer also resonated with the influence of jazz. Of course, one of the hallmarks of jazz is improvisation by the performers, thereby diminishing the composer's direct control over the music. The more that composers adopted ideas and techniques from jazz – a significant trend in the twentieth century – the more receptive they became to music that was not totally determined by their own deliberate choices.

Mathematical developments also drove trends in random music. The theory of probability had advanced quite a bit since its founding in the seventeenth century. Many discoveries in this field occurred after the time of Kirnberger's dice game, and some composers explicitly incorporated several of these new ideas into their works. Information theory, as laid out by Claude Shannon, also inspired a number of musical experiments. And the advent of the electronic computer provided a convenient means of producing a large quantity of random numbers – much handier than spending all day rolling a die!

So let's look in detail at some random music of the mid-1900s and at some of the mathematics related to it.

One Note at a Time

As we have seen, complete measures formed the building blocks of the music of the eighteenth-century dice games. Many pieces from the twentieth century, on the other hand, are based on note-by-note random choices. To take a simple example, you can generate a sequence of notes, each time letting all twelve pitch classes have equal likelihood. You could accomplish this by, say, taking twelve index cards and labeling each with a different pitch class. Shuffle the cards and pick one out at random for your first note. Shuffle all twelve again and pick a second note, and so on. (Notice that you need to put your chosen card back in the deck before selecting the next card, if you want all pitch classes to be equally likely each time.)

A procedure very much like this was used in composing the *Illiac Suite,* a piece for string quartet. In the mid-1950s, musician Lejaren Hiller and mathematician Leonard Isaacson collaborated to create this suite. It was the first extended musical composition in whose genesis computers played a significant role. Indeed, the title derives from the name of the computer Hiller and Isaacson employed. They experimented with various schemes involving randomness to produce music and then assembled some of the results into a suite.

The simplest Hiller–Isaacson scheme resembles our example above, except that instead of using the twelve pitch classes, they used thirty-one pitches spanning $2\frac{1}{2}$ octaves from a C up to an F♯. And the Illiac computer, not index cards, provided the random selection. But the basic idea remains the same: construct the melody a note at a time via a sequence of random choices, each such choice drawn from a fixed pool of equally likely alternatives. A sample of what they obtained appears as Track 21 of the CD. (Sidebar 2 comments on randomizing aspects of music other than melody, as exemplified by John Cage's work.)

Not surprisingly, these bare-bones methods result in music that most people would find of little interest. The extreme lack of structure in the note selection leads to tunes that are too – well, random. Remember the dice games, though. Their use of randomness notwithstanding, they didn't sound random because the composers put constraints on the alternatives available at each die roll. Similarly,

S I D E B A R 2

For convenience, I am restricting the discussion to the construction of melodic lines. In fact, Hiller and Isaacson also determined rhythm and dynamics randomly. For details, see their book *Experimental Music*. John Cage went even further in composing his *Music for Changes* for piano. He interpreted coin flips according to twenty-six different charts, which specified not only the pitches on a note-by-note basis, but also tempo, durations of the notes, rests, volume levels and accents, timbre, and how many notes are sounded at a given moment. There is a certain irony in this. Despite the random selection, Cage followed a highly methodical procedure and specified the music as rigidly as any serial composer.

even with note-by-note choices, you can still impose restrictions on those choices and thereby help shape the piece of music.

One such restriction is not to play with a full deck. For example, take the twelve index cards again, but now separate them into two decks: one consisting of the five notes C♯, D♯, F♯, G♯, A♯, and the other consisting of the seven notes C, D, E, F, G, A, B. The set of five cards gives you a so-called pentatonic (five-tone) scale, the black keys on a piano. If you form a melody by repeated draws from this deck, you will obtain music that many Western listeners would say sounds like Oriental music. (However, various pentatonic scales appear in traditional music worldwide, from Java to Ethiopia to Hungary to Peru.) On the other hand, the pitch classes of the deck of seven cards, corresponding to the white keys on a piano, form the C major scale. So that deck will yield melodies that at least hint at the key of C major. Hiller and Isaacson experimented with such white-key music in the *Illiac Suite* (CD Track 22).

Now restricting the notes to those of the C major scale does not of itself firmly ground the piece in that key. C major music should also emphasize the notes of the C major chord, namely, C, E, and G, especially C. But our random scheme can accommodate this, too.

How? By stacking the deck. To make certain notes more likely to occur, just shuffle in some extra cards for those notes. You could, say, form a deck of eleven cards: three C's, two E's, two G's, and one each of D, F, A, B. Then the probability of selecting C is 3/11, that for E is 2/11, and so on.

Markoved Decks

Still, all of the procedures I've discussed so far are inflexible in the sense that every time you make a random choice, your options come from the same menu. You draw from the same twelve-card deck for each and every note. Or Illiac figuratively picks from its thirty-one-card deck every time. Naturally, this leads to fairly formless music. How can there be any progression, any sense of beginning, middle, or end, when every note is picked on the same terms as all the other notes?

This problem can be overcome by using different decks at different times. Say that you want a piece somewhat reminiscent of the trios from Haydn's dice game. In the grid for the trios (Figure 1), every possible first measure is based on the G major chord. You can mimic that by selecting the first four (for example) notes from a G-major-chord deck: G, B, D. Then the next four notes might be drawn from a different deck: C, D, F♯, A, corresponding to the D^7 chord that underlies some of Haydn's versions of the second measure. Continuing in this vein, you can combine note-by-note randomness with Haydn's harmonic scheme to generate quite palatable, if not particularly inspired, music.

But despite the use of different decks, this method still proceeds fairly rigidly. In particular, the result of one choice has no impact at all on any other choice. Surely, the music could benefit from having the selection of a note take into account the previous note or notes.

So, consider the following strategy for generating a melody. Set up twelve different decks. Which deck you use will depend on the outcome of your last draw. If the last note was a C, use the first deck; if the last note was a C♯, use the second deck, and so forth. Then, according to what you draw from that deck, select the appropriate deck for your next note, and continue this sort of chain reaction. A process like this is called a Markov chain, named after the Russian mathematician A. A. Markov (1856–1922). Markov chains feature not only in the *Illiac Suite* and other early computer explorations, but also in works by musicians such as Iannis Xenakis.

Later in this chapter, I will discuss some of these compositions, but first let me describe in more detail a small-scale musical Markov chain. This sample chain will have only three pitches and therefore only three decks. You can easily adapt this description to a chain with twelve – or any other number – pitches. But as I go along, you will see why I didn't want to describe a twelve-pitch chain in full! Say that the three pitches are middle D, G, and B. (Such a setup would be appropriate for randomly generating bugle calls.) The sample chain will proceed as follows.

> If the last note was D, use a deck with 2 D's, 5 G's, and 3 B's.
> If the last note was G, use a deck with 3 D's, 3 G's, and 4 B's.
> If the last note was a B, use a deck with 2 D's, 7 G's, and 1 B.

There is no special reason to use these particular decks. I just made them up more or less arbitrarily. Having decks of ten cards each will make some of the numbers nicer to work with, but nothing says that the decks even have to be of the same size, much less consist of exactly ten cards.

The Markov procedure specifies how the latest note should influence the choice of the next note, but it does not indicate how to pick the very first note to kick off the chain reaction. Let's begin – again, arbitrarily – with D. The second note is then selected by drawing from the deck with 2 D's, 5 G's, and 3 B's. Thus D, G, and B occur as the second note with probabilities 2/10, 5/10, and 3/10, respectively. Figure 5 shows the corresponding branching-tree diagram.

Fig. 5

Now each of the three possibilities for the second note gives rise to three more branches for the third note. But notice that in Figure 6, unlike our previous diagrams, the probabilities vary according to the pitch from which the branches originate. This, of course, simply reflects the basic idea of Markov chains that the result of one link should influence the forging of the next link.

Tree diagrams, as in the case of the dice games, may facilitate the calculation of probabilities for Markov chains. For instance, what is the probability that the melody begins like *Taps*, namely, after starting with D, it goes on with another D and then a G? Tracing the branches immediately yields the answer $0.2 \times 0.5 = 0.1$. And the probability that, with the first note D, the third note is G, regardless of the second note? Simply combine the probabilities of the sequences DDG,

Fig. 6

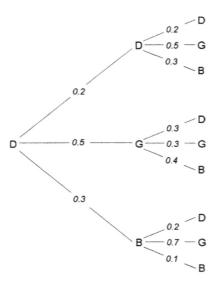

Although branching diagrams clearly indicate what goes on in a Markov chain, they can be a bit unwieldy. If Figure 6 were extended to include one more draw of a card and the decks used twelve pitches instead of three, the tree would contain $12^3 = 1728$ leaves! Obviously, a more compact representation of Markov chains would prove helpful.

Remember that I originally described the sample chain by specifying the contents of the decks and thus the corresponding probabilities. Usually, Markov chains are represented by exhibiting the probabilities in a grid, such as

		To		
		D	G	B
	D	0.2	0.5	0.3
From	G	0.3	0.3	0.4
	B	0.2	0.7	0.1

for the sample chain. Such an array of numbers is called a *matrix*. Our previous calculation of $0.46 = (0.2 \times 0.5) + (0.5 \times 0.3) + (0.3 \times 0.7)$ arises from combining the matrix's "From D" row with its "To B" column. One can combine each row with each column this way and assemble the results into a new matrix. This process, called matrix multiplication, has a surprising number of applications in fields as diverse as anthropology, physics, and economics. Moreover, the theory of matrix multiplication is closely tied in with the theory of groups introduced in Chapter 4. For further matrix information, see the book *Introduction to Finite Mathematics* by Kemeny, Snell, and Thompson.

DGG, and DBG: $(0.2 \times 0.5) + (0.5 \times 0.3) + (0.3 \times 0.7) = 0.46$. Notice how this computation follows the same lines as our previous determination of the probability of rolling 8 with a pair of dice. Techniques other than tree diagrams often help in more complex situations (see Sidebar 3).

Chained Melodies

Finally, let's take a brief look at some of the ways in which people have used and expanded on the basic Markov-chain idea to devise pieces of music. Naturally, the branch probabilities – how the decks are stacked – strongly influences the sound of the results. Hiller and Isaacson tried out several strategies for selecting these probabilities based on two factors: how well the pitches involved would blend harmonically and how close to each other the pitches would lie (CD Track 23). On the other hand, F. P. Brooks and his colleagues at Harvard took a different tack. These researchers analyzed a collection of thirty-seven hymn tunes and calculated the probability that, say, a C in those hymns would be immediately followed by an E. They then used that number for the probability of their C-to-E branch. Thus, their melodies to a certain extent reflected the properties of a preexistent body of music.

The Harvard team also experimented with letting probabilities depend on the last several notes instead of just the last one. For example, they specified the probability that the next note is F under the condition that the last three notes were CED. Again, they assigned these probabilities on the basis of their hymn sample. Hiller and Isaacson ran trials with similar multi-note considerations (CD Track 24). In addition, they combined different Markov chains by using a short chain process to generate notes played in between successive notes of the main chain.

The composer most noted for Markovian music is Iannis Xenakis. In fact, Xenakis drew inspiration from mathematics for much of his work. (Unfortunately, his writings about the mathematical aspects of his music tend to obscure, rather than clarify, his methods.) Many of Xenakis' pieces relate to probabilistic ideas, and, in particular, his *Analogique A* and *Analogique B* derive from Markov chains. In these compositions he employed chains in connection with not only pitch, but also volume and even texture – how many or how few notes appear at each instant.

In many ways, both musically and mathematically, these twentieth-century works are far removed from the dice games of Kirnberger and his contemporaries. But they all represent attempts to harness the unpredictable and blend structure with randomness. Just as a kaleidoscope builds a symmetric pattern out of a chance arrangement of bits of glass, so too may composers combine the orderly and the random in their creations.

SEVEN

PATTERN, PATTERN, PATTERN

Music lends itself well to mathematical treatment, in large part, because of all the structure inherent in music. Such structure may show up at various levels. Specifically, we can sometimes find a wheels-within-wheels effect, with a pattern that simultaneously operates in the small and in the large – and maybe in-between. In this chapter, I'll discuss a couple of comparatively recent discoveries that people have made about multilevel patterns in music.

Progressive Tendencies

As of this writing, Kevin Hamlen is a Ph.D. student in computer science at Cornell University with expected graduation date 2006. Several years ago, while still in high school, he embarked on a mathematical/musical research project. His intent was to discern, in pieces of popular music, categorizable patterns that correlated with the pieces' success. Of course, many melodic, harmonic, and rhythmic factors come into play. Hamlen focused on the sequence of chords that determines the harmonic structure of a song.

Take, for example, this simple arrangement of *Twinkle, Twinkle, Little Star* in the key of C (CD Track 25 and Figure 1).

Fig. 1

The accompaniment here consists of a series of so-called major triads, each comprising a base (as well as bass!) note and two notes at respective intervals of a third and a fifth above the base. The triad chord is named after the base note. So, in this selection the following series of triads occurs: C–F–C–G–C–G–C. Notice that the harmonies go back and forth between C chords and chords based on F and G. Such harmonic sequences are quite common – not surprisingly, in view of the importance of fourths and fifths. Indeed, the F triad is just the C triad transposed upwards by a fourth, and similarly for the G triad. The use of these common chord transitions (usually called *progressions*) makes for pleasant enough, but somewhat ho-hum, harmonies.

By way of contrast, consider *The Sound of Music.* If that song were written in the key of C as I wrote *Twinkle, Twinkle, Little Star,* its accompaniment would also begin with a C chord. That C persists throughout "The hills are alive with the sound of." But on the word "music," the harmony changes to a B♭ chord. This much less common progression lends "music" extra emphasis. Although *The Sound of Music* is not really one of my favorite things, I will admit that the C–B♭ progression works quite effectively in that opening passage.

In his analysis, Hamlen classified each chord-to-chord progression in a song as one that would "usually," "occasionally," or "rarely" occur. As you would expect, a reliance on the common progressions promotes a predictable, if not downright boring, harmonic scheme. However, there's more to the story. A steady diet of unusual progressions won't necessarily work well, either. Taken to its extreme, that route leads to a lack of structure in the harmonies,

SIDEBAR 1

To treat the rareness of a progression numerically, Hamlen assigned it a score of 1 for rare, 1/2 for occasional, and 0 for usual. (As he pointed out, this coarse-grained classification could be refined by compiling more detailed information on the relative likelihoods of progressions.) He then measured the progression-to-progression variation by taking the difference between successive scores. Thus, a switch between a rare and a usual progression gives a difference of ± 1, while following a progression by one of the same type yields a difference of 0. Finally, to indicate how much these differences fluctuate, Hamlen computed their variance. As its name suggests, the variance constitutes a statistical measure of how widely a set of numbers spreads apart. Its square root is the more well-known standard deviation. A piece of music with all progressions of the same type would have a variance of 0. If the progressions strictly alternated between usual and rare, the variance would be close to 1. Any other piece of music would fall between these two extremes. Hamlen concluded, from the songs he investigated, that a variance of around 0.4 works best.

like the unconstrained randomness we saw in the previous chapter. Hamlen concluded that "beauty . . . results from frequent and rapid changes between usual and rare progressions." So not only do the chord-to-chord progressions matter, but also progression-to-progression variations.

In fact, Hamlen extended this analysis yet another level. A steady alternation of usual, rare, usual, rare, leaves something to be desired, as well. We need more variation in the variation, so to speak. Sidebar 1 contains mathematical details of Hamlen's work, but he has illustrated the underlying ideas with a couple of songs written by the same person. Each of the graphs in Figures 2 and 3 charts the progressions over the course of a song. The high points correspond to rare progressions, the low points to usual progressions, and the intermediate–height points to occasional progressions. As

Fig. 2

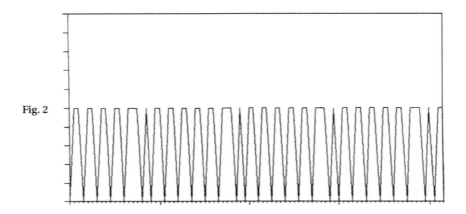

you can see from Figure 2, no rare progressions appear in the first song. Furthermore, the alternation between usual and occasional progressions exhibits fairly monotonous regularity. Figure 3 displays much more visual interest. The first fifteen progressions follow a pattern like that of Figure 2, but the usual–occasional alternation is broken up by a single spike to a rare progression. And then the graph erupts in a burst of complexity for the next twenty-five progressions before reverting to the opening scheme.

The songs? Paul Simon's *Flowers Never Bend with the Rainfall* and *Bridge Over Troubled Water,* respectively.

Fig. 3

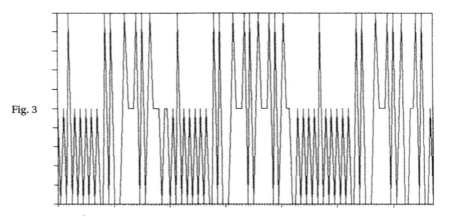

Wavy Melodies

In Hamlen's analysis, different levels of variations played a key role – variation from chord to chord, then variations in how those variations play out, and finally "third-level" variations. About fifteen years earlier than Hamlen, physicist Richard Voss, along with John Clarke, had analyzed melody, rather than harmony. The approach and techniques of Voss and Clarke differed considerably from those of Hamlen. Yet there arose the same phenomenon of patterns replicating at different levels – scales at different scales, you might say.

Think of a melodic line as tracing out a curve. The curve rises for higher notes and drops for lower notes. More or less, just play dot-to-dot with the usual musical notation (Figure 4).

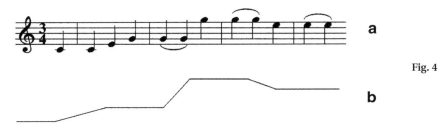

a

Fig. 4

b

We had seen in Chapter 2 that we may break a curve up into component sine curves. Back then, we had done so in the context of decomposing a sound wave that yielded a single note. But we may apply the same techniques to the curve representing a song's whole melodic pattern. With a sound wave, we had a repeating pattern. The component sine waves then repeated themselves with frequencies that were whole-number multiples of the note's frequency. In the case of a melody curve, we typically don't have a repeating pattern. As a result, the component sine curves come in a variety of generally unrelated frequencies. Each such sine curve represents a tendency of the melody to exhibit a pattern at a particular frequency. The relative contribution of a sine curve to the total curve then indicates the relative presence or absence of melodic structure on a particular time

Fig. 5

scale. (Disclaimer: To convey the main ideas, I am simplifying the goings-on at risk of distorting them. If you have some background in math, physics, or engineering, I recommend that you read the original Voss–Clarke paper for the full picture.)

For example, let's revisit *Twinkle, Twinkle, Little Star* (see Figure 5.) This piece displays patterns of repetition at various time scales. At the smallest scale, the melody consists of notes repeated in pairs. On a larger scale, measures 7 and 8 repeat measures 5 and 6. And on a still larger scale, measures 9–12 repeat measures 1–4. The Voss–Clarke calculations would highlight these three time frames. In contrast, if we generated a totally random sequence of notes, there would be no particular pattern at any time scale. For such a sequence, no time scale would be singled out by the Voss–Clarke analysis. As simple-minded as it is, *Twinkle, Twinkle, Little Star* is more interesting than total randomness, and the structure revealed through the sine curves reflects that.

A Reciprocal Relationship

Of course, for the above examples, we don't really need any mathematical apparatus to tell what's going on. But Voss and Clarke discovered something intriguing when they considered more complex

Fig. 6

musical compositions. They found a simple proportionality relationship between the relative contribution of a melodic sine component
and its period, the length of time before the pattern starts to repeat
itself.

Thus, when combining the sine components to produce the overall melodic curve as we did in Chapter 2, the sine curve with period
6 seconds would tend to be twice as high as the one with period
3 seconds. Not that music rigidly adheres to this proportionality.
This represents a tendency, rather than a strict rule. But a variety of
individual compositions, as well as the overall outputs from radio
stations featuring different genres (classical, jazz, rock), followed the
proportionality fairly closely, especially over periods between one
and ten seconds.

Voss and Clarke described their findings in terms of frequency,
instead of period. A sine curve with a period of 10 seconds repeats
$1/10$ of itself every second and so has a frequency of $1/10$ Hertz. In
general, the period is the reciprocal of the frequency f. Accordingly,
Voss and Clarke concluded that the relative contributions of sine
components (technically known as the *spectral density*) followed a
$1/f$ rule. Figure 6, taken from their paper, displays a sample of a curve
whose spectral density possesses $1/f$ behavior. (This curve did not
come from any piece of music.)

Other rules for spectral density may be used. In Figure 7, Voss
and Clarke exhibited curves with $1/f^2$ and $1/f^0$ rules, respectively,
which convey quite different visual impressions. Since $1/f^0 = 1$, regardless of f, the $1/f^0$ rule puts all time scales on an equal footing. As
mentioned previously, this occurs with a totally random, structure-
less fluctuation. The curve bounces up and down with no regard for

$1/f^2$

Fig. 7

$1/f^0$

what has come before. (In fact, this is the same curve that appeared in the context of white noise in Chapter 2, where it represented random variations in air pressure, rather than in melody.)

On the other hand, the $1/f^2$ curve evinces quite a good "memory." As it progresses from left to right, its height changes by small random amounts, so that for long stretches the height remains at roughly the same level. The $1/f$ curve's shape compromises between the two forms in Figure 7. I'll let you decide which formula produces more visual interest, but again, Voss and Clarke found that music generally resembles Figure 6.

On Closer Inspection

How does all of this relate to my original point of departure, the replication of patterns at various levels? That idea is already implicit in the decomposition of the melodic curve into copies of the sine wave that differ in their time scales. But there's more. Whenever a curve's spectral density behaves like $1/f^p$ for some power p, the curve exhibits so-called self-similarity. This means that if we were to take a fragment of one of the shapes in Figures 6 or 7 and magnify the fragment, we would see a pattern like the one we started with. The ups and downs of any little piece of the picture reflect, in miniature, those of the overall picture. And if we took a tiny bit of the smaller piece and blew it up, the pattern would persist. A full-fledged $1/f$ curve preserves the same kind of outline, no matter how strong the magnifying glass we use to inspect it. Similarly, a true $1/f^2$ curve appears $1/f^2$ at all scales.

No piece of music is exactly $1/f$. Again, the $1/f$ behavior typically holds only over a certain range of frequencies, and even then it will not hold precisely. In fact, a completely $1/f$ (or $1/f^2$, etc.) curve forms a fractal, which can never be totally realized physically. But the approximate self-similarity implied by $1/f$-like spectral density

Voss and Clarke generated music of different spectral densities by using appropriate electronic circuitry. When Martin Gardner reported on $1/f$ music for his *Scientific American* column in 1978, he asked Voss for an easier method that Gardner's readers could use at home. Voss came up with the following procedure, which although not strictly $1/f$, comes reasonably close.

With three dice you can obtain totals between 3 and 18. (Some of these totals will come up more often than others; see Chapter 6.) Assign a different note to each total. Now take three dice of different colors, say, red, blue, and yellow. Roll the dice and record the total. Then, leaving the blue and red alone, re-roll the yellow and record the new total. Next, leaving the blue alone, re-roll both the red and yellow and record this third total. Finally, re-roll just the yellow for a fourth total. Now repeat this, rolling all three dice and re-rolling as above to crank out another four totals, and so on. When you are done, convert the totals to their assigned notes.

In this procedure, you roll the yellow with every new total, while you roll the red only every other time, and the blue every fourth time. The different time scales at which the dice operate lend the results their $1/f$-like characteristics.

SIDEBAR 2

Fig. 8

indicates that music tends to display patterns that are mirrored at different levels.

So far, we've been looking at the analysis of standard musical selections. One can, of course, devise music specifically based on

$1/f$ rules. This provides a twist on the random compositions of the previous chapter. There we considered how composers avoided total randomness by constraining the options, such as precomposed measures that adhered to a certain harmonic scheme. By generating random $1/f^p$ music, one does not explicitly impose any particular structure. Rather, the structure comes implicitly from the nature of the spectral characteristics. Voss and Clarke experimented with this as a compositional approach. (See Sidebar 2 for instructions on how you can easily create $1/f$-like music.) Figure 8 displays some of their results. In these pieces, not only the pitches, but also the rhythms, come from $1/f^p$ rules. As expected, the $1/f^0$ music skips up and down fairly wildly, while the $1/f^2$ melodic line remains fairly static, inching its way along for the most part. The $1/f$ excerpt, lying between those extremes, follows contours that visually resemble typical music. If you listen to this passage's initial notes on CD Track 26, however, you are not likely to walk away humming the tune. Since the late 1970s, when Voss and Clarke published their work, some musicians have elaborated on such experiments with $1/f$ rules. But between these deliberate uses of multilevel patterning and Paul Simon's more intuitive uses of the same concept, there's no contest, at least on the sales charts.

EIGHT

SIGHT MEETS SOUND

Of the different branches of mathematics, geometry allies itself most naturally with the visual arts. The use of perspective in drawing and painting, based on geometric principles, represented a major development in the history of art. Geometry's role in music seems less obvious. But musical notation tries to depict auditory phenomena in spatial terms, and people have exploited this eye–ear link in various ways. Music also combines with spatial activity whenever dancing occurs. This chapter will sample some of the uses of geometric ideas in musical contexts.

Ups and Downs…

In previous chapters we have taken the phrase "melodic line" literally, using the rises and falls of a tune to trace out a graph. Why not reverse the process? Given a curve, we can superimpose it on a musical staff and construct a melody to follow that outline. Not just a plaything of mathematicians or avant-gardists, this strategy found its way into the repertoire of mainstream composers.

The reputation of Heitor Villa-Lobos (1887–1959) rests mainly on his many pieces that reflect the musical heritage of his native Brazil. But he also composed works by a method he called millimetrization, which boils down to the above-mentioned strategy. Villa-Lobos

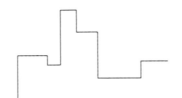

Fig. 1

would start with a photograph. For his Sixth Symphony, he used the mountains of Rio de Janeiro. In another piece for piano/orchestra, he depicted the New York City skyline. Whatever the source of the contour, he would render it on graph paper. Then, in the words of Eero Tarasti, "the most important points ... of the profile are indicated. The points correspond to the tones to be chosen."

As Tarasti remarked, "the method leaves its adapter to ponder which dots in the photographic profile are ... turning points in the melody." A skyline such as that in Figure 1 lends itself to a fairly cut-and-dried interpretation. You can immediately translate this into a sequence of notes, letting the length and height of each horizontal segment determine, respectively, the length and pitch of the corresponding note. But what about a mountain or even just a semicircle? You need to decide how to discretize the continuous curve to obtain a tune, especially for instruments like pianos that do not play a continuous range of pitches. (In his Sixth Symphony, Villa-Lobos did take advantage of the string section's ability to provide a continuous glissando.) Tarasti credits Villa-Lobos with interpreting the graph in light of his compositional skills in order to achieve certain results.

Serge Prokofiev has been reported to employ a similar technique. According to a correspondent cited by Martin Gardner, Prokofiev converted outlines into melodies in his film score for *Aleksandr Nevsky.* Supposedly, Prokofiev took stills from the footage for various scenes and based musical themes on the visual contours. Prokofiev and the movie's director Sergei Eisenstein did collaborate closely, with the filming and the scoring exerting reciprocal influences on each other, but I have not yet been able to confirm the specific claim that Gardner reported.

Techniques resembling millimetrization also played an important role for Joseph Schillinger (1895–1943). Schillinger composed many works, including the *First Airphonic Suite* (1929), for theremin – an early electronic instrument – and orchestra. He is best remembered, however, as a music theorist and composition teacher. Schillinger devised a compositional system purportedly based on mathematical principles, although I must say that I have found his writings fairly opaque. Still, in his heyday he was quite influential. George Gershwin studied under Schillinger and used his system in writing *Porgy and Bess*.

SIDEBAR 1

The millimetrization scheme is a natural one for skylines and mountains, which have an intrinsic up-and-down orientation. However, for other contours this method can seem somewhat arbitrary. Take the outline of a snowflake, for instance. It has no single preferred orientation. If you rotate it around its center, it recognizably retains its shape. But the resultant millimetrized melody could differ considerably. This problem shows up vividly if we consider a single line segment. Turn it at a different angle, and you still have a line segment of the same length (Figure 2). The two segments might not produce the same music, but we could reasonably hope that their corresponding tunes would at least share a family resemblance. Instead, millimetrization yields melodies that have little likeness. The

Fig. 2

one sustains a near-monotone for a long stretch of time, while the other very rapidly covers a wide range of pitches.

Here's an alternative approach that avoids this drawback. Rather than scanning the contour from left to right as you observe it from above, trace your way around the contour. At each moment, sound a note that corresponds to the direction in which you are moving. In terms of the snowflake example, you would chart a path around the snowflake's edges, changing notes each time you turn a corner. Notice that any line segment generates a single note. The length of the note will vary according to the segment's length, and the pitch will vary according to the segment's orientation. Of course, with circular arcs and the like, you face the same matters of discretizing the continuous as with millimetrization. I devised this scheme and experimented with it several years ago. The results? I got some interesting passages, but let's just say that I'm no Villa-Lobos!

...Twists and Turns

In 1986, Przemyslaw Prusinkiewicz introduced yet another means of converting contours to music. This method, later elaborated upon by Stephanie Mason and others, shares some features with both millimetrization and my scheme, and it contains some twists of its own. To start with, Prusinkiewicz's method only applies to pictures consisting solely of horizontal and vertical line segments. Again, city skylines lend themselves well. But so does something like Figure 3 (more about this particular curve in a little while). Follow the path

Fig. 3

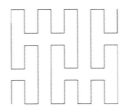

of Figure 3 from one end to the other. When you do so, you will alternate between horizontal and vertical segments. Each successive
horizontal–vertical pair will determine a note for you. Not too surprisingly, the length of the pair's horizontal segment will correspond
to the length of the note. The length of the pair's vertical segment
will correspond to the *change* in pitch from the previous note. For
example, using semitone steps, a vertical climb of three units would
result in a note three semitones higher than the one before. A vertical descent similarly would lower the pitch. If we assume an initial
base-level pitch of C, Figure 3 generates the following melody (CD
Track 27 and Figure 4).

Fig. 4

Where did that contour come from in the first place? Forgetting
the little zigs and zags, the overall track looks like Figure 5.

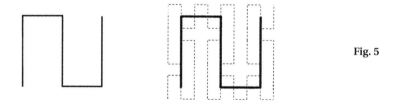

Fig. 5

But the zigs and zags themselves consist of little copies of that same
overall shape, along with a few connecting segments (Figure 6).

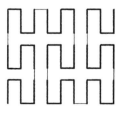

Fig. 6

109

In other words, we have the self-similarity phenomenon of the previous chapter, where a pattern appears simultaneously at large and small scale. We could add a third level to this process (Figure 7).

Fig. 7

Here, copies of Figure 5 are strung together, following the path of Figure 3. (Or with equal justification, we could say that copies of Figure 3 are strung along the path of Figure 5.)

Alternatively, we can describe these curves in a nonpictorial fashion by specifying the sequence of turns. The iterative process that generates the successive curves can then be concisely presented in terms of so-called L-systems (Sidebar 2). For more details, other curves, and further references, read the paper (see Bibliography) by Stephanie Mason and Michael Saffle. Incidentally, if we continued this iteration forever, we would obtain a fully self-similar pattern. The resulting curve has the remarkable property of passing through

In 1968, Aristid Lindenmayer mathematically modeled developmental processes in biology. His models have since come to be known as Lindenmayer systems, L-systems for short. An L-system consists of a starting string of letters, along with rules for replacing each letter with another string of letters. For example, we could use the rules

A gets replaced by *ADBDARDRBDADBLDLADBDA*;

B gets replaced by *BDADBLDLADBDARDRBDADB*;

D, *L*, *R* get replaced by themselves.

If we start with just the letter *A*, then for the second stage, we replace that *A* with the string *ADBDARDRBDADBLDLADBDA*. For the third stage, we replace every second-stage letter according to the rules. Thus, we obtain a very long string that begins

ADBDARDRBDADBLDLADBDA D BDAD . . .

We can convert the string at each stage into a picture by interpreting *D* as "draw a line segment (of specified length) in the direction we're facing," *L* as "turn left by 90°," and *R* as "turn right by 90°." For purposes of the picture, we ignore the *A*'s and *B*'s. The second-stage string *ADBDARDRBDADBLDLADBDA* yields Figure 5.

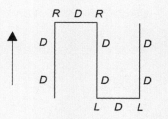

Likewise, the third-stage string produces Figure 3, and so on.

every point of the square. By coincidence, the curve bears the almost-musical name of the mathematician who introduced it, Giuseppe Peano.

Bell-y Dancing

In a totally different vein, music and geometry naturally meet on the dance floor. Of the many kinds of dancing, I will focus here on the contradance for a couple of reasons. First, some mathematicians have very recently applied their professional skills to this dance form, not just to analyze it but also to create additions to the contradance repertoire. Second, we have already dealt with the relevant math, namely, group theory (see Chapters 4 and 5). Nonmathematicians often associate geometric studies with the Euclid-styled proofs about triangles typical of the traditional high-school curriculum. But geometry as practiced by mathematicians and applied by physicists actually uses groups to a great extent. We will have a little taste of that here.

SIDEBAR 3

Both the name *contradance* and the dance style itself trace back to old English country (i.e., folk) dances. Around the eighteenth century, the French built on the English format and called the result *contredanse*. The French name translates not as "country dance," but rather as "counter-dance," in reference to the motion of couples counter to each other. This nomenclature then found its way back into English as *contradance*, as well as into German as *Kontretanz*. Beethoven composed a set of twelve *Kontretanz* for orchestra. The seventh in this set employed a theme that is more well-known for its appearance in Beethoven's Third Symphony, his *Eroica Variations* for piano, and his *Creatures of Prometheus* ballet.

In contradancing, couples form a line with partners facing each other, as in Figure 8.

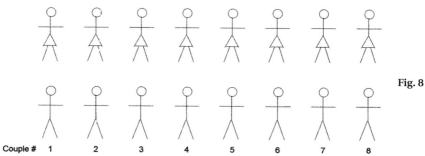

Couple #   1   2   3   4   5   6   7   8

Fig. 8

As the dance unfolds, couples trade positions with neighboring couples. Thus, after a certain amount of footwork, the couples will change to the following configuration: 2 1 4 3 6 5 8 7. This might ring a bell, figuratively and literally. The swapping of positions by adjacent couples is just like the swapping that occurs in the sequence of bells in change-ringing. In fact, the contra pattern continues like the four-bell routine in Chapter 5 (Figure 9).

$$1 \; 2 \; 3 \; 4 \; 5 \; 6 \; 7 \; 8$$
$$\updownarrow \quad \updownarrow \quad \updownarrow \quad \updownarrow$$
$$2 \; 1 \; 4 \; 3 \; 6 \; 5 \; 8 \; 7$$
$$\updownarrow \quad \updownarrow \quad \updownarrow$$
$$2 \; 4 \; 1 \; 6 \; 3 \; 8 \; 5 \; 7$$
$$\updownarrow \quad \updownarrow \quad \updownarrow \quad \updownarrow$$
$$4 \; 2 \; 6 \; 1 \; 8 \; 3 \; 7 \; 5$$
$$\updownarrow \quad \updownarrow \quad \updownarrow$$
$$4 \; 6 \; 2 \; 8 \; 1 \; 7 \; 3 \; 5$$
$$\vdots$$

Fig. 9

113

Unlike the change-ringing scheme, however, this swapping pattern continues throughout the whole contra. Remember that with bells, we needed to break from this pattern occasionally to ensure that all possible orders of the bells came up. In a contra with eight couples, the dancers – wisely – do not attempt to arrange themselves in all 40,320 orders. They will pass through only sixteen different configurations before returning to their starting positions.

I mentioned earlier that the couples change the configuration after some footwork. The dance would be very dull if it consisted of nothing other than couples walking to their new positions. Rather, each couple goes through a series of actions with the couple whose place they are about to take over. For example, the foursome may join hands and rotate their little square a quarter-turn (see Figure 10).

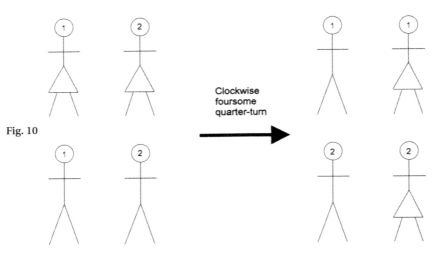

Fig. 10

Then within each couple, the partners may face each other and do a half-turn, exchanging positions (Figure 11).

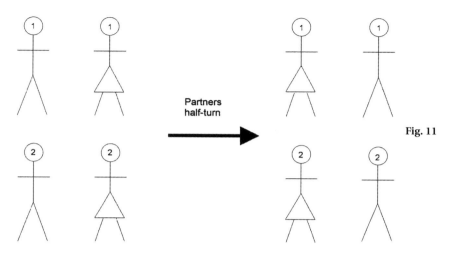

Partners
half-turn

Fig. 11

Another foursome quarter-turn then accomplishes the exchange of the couples' locations (Figure 12).

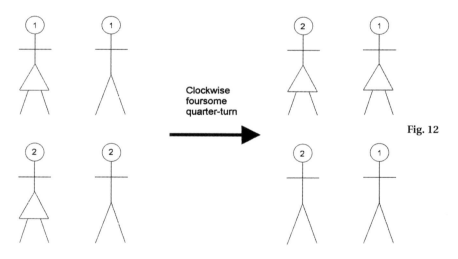

Clockwise
foursome
quarter-turn

Fig. 12

The quarter-turn behaves like the transposition T_3 of Chapter 4. If you perform it four times, you end up the same as you started. And the partner half-turn actually mirror reflects the conformation of the little square. Such a reflection combines with the quarter-turn in the

115

same way that the inversion I combines with transposition. We had seen previously that $T_3 I T_3 = I$. Likewise, a quarter-turn followed by a reflection followed by another quarter-turn gives a net result of a reflection, swapping the couples as desired.

Actual contradances exhibit more complexity than this simple example. However, the sequence of steps still must ultimately result in the reflection that trades positions, one couple with another. Mathematicians such as Bernie Scanlon in California and Larry Copes in Minnesota have designed new contras in recent years, explicitly using the underlying group-theoretic geometry. As with the schemes we saw earlier in this chapter, the mathematics of space serves to help create new works in the musical realm.

NINE

HOW *NOT* TO MIX MATHEMATICS AND MUSIC

In previous chapters we have seen the power of mathematics in helping us better understand various aspects of music. Now let's look at some examples of failed attempts at combining the two fields. These misfires, often interesting or even amusing in their own right, also serve as reminders to exercise caution around attempts to analyze music mathematically. Not surprisingly, many people run aground in such attempts because of their shaky grasp of mathematics. But I'll start with a case study that shows that even the best of mathematicians can go astray.

Aesthetic Arithmetic

George David Birkhoff ranks as one of the top American mathematicians of the first part of the twentieth century, and his work remains highly influential to this day. His more than 200 publications span a wide range of mathematical areas. Indeed, Birkhoff helped to found and shape several of these areas. Moreover, many of the forty-five doctoral students whose dissertations he supervised became the stars of the next generation. Birkhoff was unquestionably a great mathematician.

As a mathematician of much more modest credentials, I therefore feel somewhat hubristic here. G. D. Birkhoff as an example – though certainly not the worst – of how not to mix mathematics and

In his paper *Polygonal Forms* Birkhoff rated ninety straight-line figures according to his aesthetic measure. You might want to rank them yourself and/or predict what Birkhoff said before looking at his scores in Sidebar 2.

music? I'm afraid so. In fact, part of why Birkhoff's attempt is such a letdown comes from a sense that he should have done a much better job of it, given his abilities and his enthusiasm for this topic.

Birkhoff actually pursued a more ambitious undertaking: to develop a general mathematical theory of aesthetics and apply it to works of visual art (see Sidebar 1) and literature, as well as music. Over a period of many years beginning in the late 1920s, he expounded his ideas on aesthetics in several publications and lectures. Birkhoff engaged the interest and enlisted the aid of his Harvard colleagues. Among these was no less a figure than Walter Piston, the author of classic textbooks on musical composition and two-time winner of the Pulitzer Prize.

So what was Birkhoff's theory? It all boiled down to assessing a work's aesthetic value by the formula O/C. Here O is not zero, which would really make for a simple theory! Rather, O denotes the amount of Order present in the work, while C represents the amount of Complexity. In Birkhoff's view, something's aesthetic attributes arise from the harmonious relationship of its components to each other. The term O measures this internal structure. However, to Birkhoff the amount of effort required to perceive the work in the first place will detract from the appeal of its structure. The term C reflects this effort, and the greater the value of C in the formula, the smaller the aesthetic value.

Certainly Birkhoff's premises are debatable at a qualitative, much less quantitative, level. Does complexity really diminish artistic worth? Doesn't a large part of the visual appeal of snowflakes or fractal pictures stem precisely from their intricacy? And why the ratio O/C? Why not $O - C$ or O/C^2 or ... ? Birkhoff likened his aesthetic measure to the rate of return on an investment, likewise computed by a ratio, but admitted that his justifications were not necessarily convincing.

An Orderly Gauge of Complexity, A Complex Gauge of Order

Still, in and of themselves, the above shortcomings do not constitute fatal flaws in Birkhoff's theory. One would not expect a first attempt at such a grand project to capture all the contingencies and with pinpoint accuracy. But one should expect the theory to supply a useful conceptual framework that simplifies and unifies by highlighting general underlying principles. With the concise gravitational formula $F = GmM/r^2$, Newton described a fundamental property of matter that accounts for behaviors as apparently different as a fig falling from a tree and a planet orbiting a star.

Now Birkhoff's O/C seems to have similar potential for getting to the heart of the matter. It is as straightforward a formula as we could desire. And it isolates two basic properties of a work of art. The disappointment lies in how Birkhoff assigned a value to O for a piece of music, that is, how he determined how much pattern underlies the piece. C poses no real problem. He assessed a melody's complexity by how many notes it contains – a very crude, but not unreasonable criterion. On the other hand, Birkhoff's determination of O lacks the directness of his C or of the quantities like mass and distance that appear in Newton's formula. Nor does it rest on simple, unified principles. Rather, Birkhoff concocted a convoluted scoring system that somewhat arbitrarily awards or removes points according to a seemingly ad hoc collection of musical rules.

In his monograph, *A Mathematical Theory of Aesthetics*, Birkhoff listed nineteen different categories in his points system for melodies, with a side comment that this list was almost surely incomplete. All that even before taking harmony and rhythm into account! The following excerpts indicate his approach: "(5) Repetition in Comparable Phrases: If a part of one phrase at least a measure in length is repeated in a later phrase with which it is compared, or if notes in corresponding position are repeated, there is a count of 1 [point] for each repeated note. [. . .] (11) Harmonic Contrast: If all of the notes of a measure but one fall in a consonant major chord there is a count of 1 [point] each for the last two notes."

In effect, Birkhoff said that a piece of music rates as good by music textbook standards if it follows the rules given in the music textbooks. This logic is unassailable, but hardly informative! And that's the trouble with Birkhoff's scheme. Ultimately, it provides no new insights. Rather than the clarification and unification that Newton provided with his formula, Birkhoff just transferred the dirty work of assessing aesthetic value to that of evaluating the quantity O, while offering nothing more than a motley collection of rules of thumb to do the latter.

I do not feel that Birkhoff's attempt was doomed to failure from the start. Personally, I have no philosophical objections at all to the idea of a mathematical formula that measures beauty. But I will admit that I have yet to see one I consider useful. Certainly, Birkhoff's doesn't do the job, despite his obviously earnest intentions and great talents.

Hare-Raising Numbers

Some of the most misguided attempts to link music and mathematics have involved the Fibonacci numbers and the related golden ratio (see Sidebar 3). These numbers possess various interesting

> According to Birkhoff, believe it or not, the figures decrease in aesthetic value, reading from left to right. He assigned them respective scores of 1, 0.9, 0.58, 0.43, 0.25, 0.17, 0, and −0.11.

SIDEBAR 2

mathematical properties and sometimes appear in nature, especially in patterns of things such as sunflower seeds and daisy petals. But this does not suffice for some people. A body of literature exists that ascribes not just mathematical, but also mystical, attributes to these numbers and finds them not just in certain patterns of plant growth, but everywhere, even in places where they aren't.

Take this example, which appears in a book published by a reputable company specializing in educational materials. The book claims that Fibonacci numbers appear, among other places, in the lengths of musical scales:

Pentatonic scale	C D E G A	5 notes
Diatonic scale	C D E F G A B C	8 notes
Chromatic scale	C C♯ D D♯ E F F♯ G G♯ A A♯ B C	13 notes

I will look at this example in some detail, because it encapsulates many of the flaws seen in musical numerology. Among several fallacies here, the dubious method of counting notes probably jumps out most obviously. The pentatonic scale, indeed, has five different notes to it. But the diatonic possesses seven, not eight; likewise, the chromatic is a twelve-tone scale. The author inflated the counts by including an extra C. Of course, if you prefer, you may end a scale an octave above where you started, but then the pentatonic scale would (despite its name) contain six notes. In any event, the author gave

In 1202, Leonardo of Pisa, also known as Fibonacci, wrote a mathematics book. One exercise in the book dealt with rabbit populations. Under the somewhat unrealistic assumptions of the exercise, the number of pairs of rabbits in successive months follows the pattern 1, 1, 2, 3, 5, 8, 13, and so on, in which two consecutive numbers sum up to yield the next one. This Fibonacci sequence of numbers has close ties to the quantity $(\sqrt{5} - 1)/2$ (approximately 0.618), which is often denoted φ. The number $1/\varphi$, which happens to equal $\varphi + 1$, is often denoted τ; both φ and τ have been called the "golden ratio." The formula $(\tau^n - (-\varphi)^n)/\sqrt{5}$ computes the nth number to appear in the Fibonacci sequence. As a consequence, the further along in this sequence you go, the closer the ratio between a Fibonacci number and its successor gets to φ. For instance, 8/13 is around 0.615. The Fibonaccis possess many properties, some just interesting curiosities, others useful in computer applications and elsewhere. As one example, the sum of the first n Fibonaccis always equals one less than the $(n + 2)$th Fibonacci.

no reason for treating the scales differently. Only such Procrustean tinkering, however, would let the author make a claim for Fibonacci numbers.

Further, the author did not mention other scales. In particular, the whole-tone scale (C, D, E, F♯, G♯, A♯) contains six notes – seven, if we allow another C at the end. Neither six nor seven happens to be a Fibonacci number, though, and so this scale did not get listed. It is clearly easier to claim that the Fibonacci numbers play a significant role in scales if you disregard those scales that don't cooperate with you.

But what if the whole-tone scale had only five notes and the other scales had also been of Fibonacci lengths without tinkering? Would that make a case for the musical significance of Fibonacci? Not necessarily. After all, of the numbers from 1 up through 13, six – nearly half of them – are Fibonacci. So, if you're dealing with a smallish number, you have a fairly decent chance for it to be Fibonacci. And even if it isn't, you can often easily shoehorn it into a Fibonacci role if you choose.

For example, my birthday happens to be April 27, 1955. Writing that in the usual form 4/27/55, I notice that $4 = 2^2$ and $27 = 3^3$, and 2 and 3 are consecutive Fibonacci numbers. Moreover, 55 consists of side-by-side copies of 5, the next Fibonacci, as well as being a Fibonacci number in its own right. Do any of these tidbits signify anything? Of course not.

The Canadian mathematician Richard Guy devoted his paper *The Strong Law of Small Numbers* to pointing out that "[t]here aren't enough small numbers to meet the many demands made of them." If you look at some numbers, you may very well find Fibonacci numbers, or square numbers, or prime numbers, or any other favorite kind of number you have. But unless you can also find a particular reason why numbers of a certain type *should* appear, you're quite likely just turning up a coincidence.

Summarizing so far, then, coincidences will happen. You can increase their occurrence by appropriate tinkering. And if you blot out of your mind the times that the number-juggling didn't work, you can convince yourself and/or others of patterns that do not really exist (see Sidebar 4). The scales example I cited is fairly easy to see through. However, another claim for the Fibonacci numbers – more sophisticated, but ultimately based on the same fallacies – has gained many adherents.

Myth and Music

A piece of folklore about mathematics and music, a sort of musical urban legend, persists to the effect that the great Hungarian composer Béla Bartók based much of his music on Fibonacci numbers and the golden ratio. Musicologists such as László Somfai have

SIDEBAR 4

For many more examples, see Underwood Dudley's book *Numerology, or, What Pythagoras Wrought*. People use the fallacies we have examined to claim omnipresence for all sorts of numbers besides the Fibonaccis, such as squares, pi, seven, thirteen, and the ever-popular 666. Dudley drolly describes such activities, ranging from an anonymous person's serious obsession with finding 37's and 73's in the Bible to Monte Zerger's tongue-in-cheek compilation of connections between Fibonacci numbers and the state of Illinois.

debunked this notion. Somfai inspected "the complete existing source material of Bartók's compositions as well as manuscripts of folk-music transcriptions, drafts of articles, and scattered scrap papers in the Hungarian and American estate." He observed that "not a single calculation of the proportions of a composition – with Fibonacci or other numbers – has been discovered," despite "the composer's notorious lifelong habit of keeping and recycling every bit of paper." But the legend still lingers.

The idea of a Bartók–Fibonacci connection traces to the musicologist Ernő Lendvai. In his writings, he presented many examples from Bartók's works, purporting to show how Fibonacci suffused that music. Interestingly, even Lendvai stopped just short of claiming that Bartók deliberately used Fibonacci numbers and the golden ratio as compositional devices. Rather, Lendvai wrote, "I have no wish to prove that he aspired to an arithmetic or geometric system; he did, however, by going back to the roots of music, discover fundamental laws and 'root' correlations which *may be expressed* [emphasis added] by formula-like, mathematical symbols." The legend has since expanded to that of Bartók, the numerological composer.

Whether by deliberate design or not, does Fibonacci recur in Bartók's music? A second look at Lendvai's examples reveals them to be as illusory as the Fibonacci scales – and for the same reasons. First, the Law of Small Numbers operates with a vengeance here. Bartók wrote a lot of music. While not as prolific as Haydn, say, Bartók composed works in a variety of genres: orchestral music, vocal music, ballets, an opera, a cantata, several concertos, six string quartets and other chamber music, and a sizeable piano literature. His six-volume piano collection *Mikrokosmos* alone consists of over 150 pieces. With such a body of music, it would be miraculous if you *couldn't* find several instances in which the count for something or other turned out as a Fibonacci number or two counts stood approximately in golden ratio.

For example, Lendvai pointed to a scale appearing in Bartók's *Cantata Profana*. This scale consists of a starting D, followed by notes at 2, 3, 5, 6, and 8 semitones above that D. Remember that a smallish number has a good chance of being Fibonacci in the first place. In this instance, not all of the numbers are Fibonacci; however, Lendvai focused on just the ones that are. He described this as a Fibonacci 2, 3, 5, 8 scale with an added diminished fifth note (the 6), thereby sweeping aside an inconvenient note that did not fit his preconceived notions. This scale could with equal justice be called a 2, 3, 6, 8 scale with a 5 added, but that interpretation doesn't match his scheme and so got ignored.

Lendvai repeatedly juggled things around to support his claims. In discussing the first movement of the *Sonata for Two Pianos and Percussion*, he counted measures to make his point. But for the first movement of the *Divertimento*, whose measures didn't cooperate, he counted triplets. He explained this away by saying "the number of bars is irrelevant owing to their variable time-signatures," although the changing time signatures of measures didn't faze him when considering the *Sonata*. And when analyzing the *Sonata* as a whole, he enumerated eighth notes instead, with no explanation of why he used a different object to count than he used for the first movement. Again, if you examine enough passages and try enough tinkers, sooner or later you'll come across a near-golden ratio. By presenting several

Musicologists often describe movements of sonatas as consisting of an *exposition*, which presents musical themes, and a longer *development and recapitulation*, in which the composer plays with the themes (using devices such as those described in Chapter 4) and then brings the movement to a close. Let a and b, respectively, represent the number of measures in these two segments. Some enthusiasts of the golden ratio φ have pointed out that in Mozart's piano sonatas, the ratio $b/(a+b)$ usually falls quite close to φ. However, John Putz has placed this in perspective. In his paper, *The Golden Section and the Piano Sonatas of Mozart*, Putz pointed out that if the movement is really in near-golden proportion, then a/b also has to approximate φ. But Mozart's a/b values do not resemble φ nearly as closely as the $b/(a+b)$ values. In fact, if $0 < a < b$, then $b/(a+b)$ will *always* lie closer to φ than a/b will. Thus by focusing on $b/(a+b)$, rather than a/b, enthusiasts automatically stack the deck in φ's favor.

such examples – and not the many, many places where Fibonacci does not show up – Lendvai compiled a superficially impressive, but ultimately meaningless, list.

In fact, one example that Lendvai presented undercuts his own case. He revisited the first movement of the *Sonata for Two Pianos and Percussion*, this time counting whole notes! Coming up with a total of 804, Lendvai found it significant that "$2^8\pi = 804$." Let us ignore the fact that this "equation" is not true – indeed, $804/2^8 = 804/256$ does not even give a particularly great approximation to pi; $355/113$ does much better. More than anything else, though, this example demonstrates vividly that Lendvai's claims rest not on any true Bartók–Fibonacci connection, but rather on Lendvai's patience and cleverness in number-juggling. Lendvai for once couldn't make 804 fit the Fibonacci mold. Instead, he expressed 804 (approximately) in terms of some other interesting numbers. All this proves is that there are lots of ways to play with numbers. If you have the time and inclination, you can find not only Fibonacci, but also pi, powers of two, prime numbers, or what have you in the music of Bartók – or in the phone book.

Let me point out that I am not accusing either Lendvai or the author of the passage on scales of deliberate fraudulence. Similar misapplications of mathematics abound (for a related example, see Sidebar 5) and very often represent self-delusion. And with good reason. The human mind possesses a wonderful and highly useful ability to perceive patterns. Sometimes our mental mechanisms push us to see patterns that aren't really there. Take a couple of pattern-rich subjects like music and mathematics, and the temptations can become irresistible. Just as with Birkhoff, our enthusiasm can get the better of us. In the end, all these failed attempts pay tribute to the beauties of music and mathematics and the fascination they exert on people.

BIBLIOGRAPHY

The materials listed below were mentioned in the text, were used as reference sources by me, or discuss relevant topics.

Chapter 1

G. Assayag, H. G. Feichtinger, and J. F. Rodrigues (editors). *Mathematics and Music*. Springer-Verlag, New York, 2002. [The contributions are of variable quality, but some are quite good. See http://www.maa.org/reviews/mathmusic.html for my *MAA Online* review of this book.]

J. Fauvel, R. Flood, and R. Wilson (editors). *Music and Mathematics*. Oxford University Press, Oxford, 2003. [Lots of good material.]

R. Smullyan. *5000 B.C. and Other Philosophical Fantasies*. St. Martin's Press, New York, 1983.

Chapters 2 and 3

J. Backus. *The Acoustical Foundations of Music*. W. W. Norton, New York, 1969.

J. Barbour. *Tuning and Temperament*. Michigan State College Press, East Lansing, MI, 1953.

BIBLIOGRAPHY

E. Blackwood. *The Structure of Recognizable Diatonic Tunings.* Princeton University Press, Princeton, NJ, 1985.

D. Butler. *The Musician's Guide to Perception and Cognition.* Schirmer Books, New York, 1992. [Includes an audio CD with musical examples.]

E. Dunne and M. McConnell. Pianos and continued fractions. *Mathematics Magazine* **72**, 104–115, 1999.

N. Fletcher and T. Rossing. *The Physics of Musical Instruments.* Springer-Verlag, New York, 1991.

D. Hall. *Musical Acoustics.* Wadsworth Publishing, Belmont, CA, 1980.

A. J. M. Houtsma, T. D. Rossing, and W. M. Wagenaars. *Auditory Demonstrations.* Acoustical Society of America, Sewickley, PA, 1989. [Audio CD]

O. Jorgensen. *Tuning.* Michigan State University Press, East Lansing, MI, 1991.

E. Neuwirth. *Musical Temperaments.* Springer-Verlag, New York, 1997. [CD-ROM]

E. Neuwirth. The mathematics of tuning music instruments – A simple toolkit for experiments. In G. Assayag, H. G. Feichtinger, and J. F. Rodrigues (editors), *Mathematics and Music.* Springer-Verlag, New York, 2002. [Includes the Mathematica code used to generate sounds for his *Musical Temperaments.*]

E. Neuwirth. Music and Fun (and some Mathematics). http://sunsite.univie.ac.at/musicfun/. [Besides material on tunings, this Web site has links to musical dice games and much more.]

M. R. Petersen. Musical analysis and synthesis in Matlab. *College Mathematics Journal* **35**, 396–401, 2004.

M. R. Petersen. Resources to accompany musical analysis and synthesis in Matlab. http://amath.colorado.edu/pub/matlab/music. [Matlab code and other material related to the above article.]

BIBLIOGRAPHY

J. Pierce. *The Science of Musical Sound.* Scientific American Books (W. H. Freeman), New York, 1983. [Freeman issued a revised second edition, but the first edition includes a vinyl disc with audio examples.]

J. Roederer. *The Physics and Psychophysics of Music.* Springer-Verlag, New York, 1995.

T. Svobodny. *Mathematical Modeling for Industry and Engineering.* Prentice-Hall, Englewood Cliffs, NJ, 1998. [Chapter 9 contains a detailed model of the workings of the human ear.]

C. Taylor. *The Physics of Musical Sounds.* English Universities Press, London, 1965. [Includes a vinyl disc with audio examples.]

Chapter 4

M. Babbitt. Set structure as a compositional determinant. *Journal of Music Theory* **5**, 72–94, 1961.

F. J. Budden. *The Fascination of Groups.* Cambridge University Press, Cambridge, 1972.

A. Forte. *The Structure of Atonal Music.* Yale University Press, New Haven, CT, 1973.

W. Hodges and R. Wilson. Musical patterns. In G. Assayag, H. G. Feichtinger, and J. F. Rodrigues (editors), *Mathematics and Music.* Springer-Verlag, New York, 2002.

P. Isihara and M. Knapp. Basic Z_{12} analysis of musical chords. *UMAP Journal* **14**, 319–348, 1993.

B. McCartin. Prelude to musical geometry. *College Mathematics Journal* **29**, 354–370, 1998.

G. Perle. *Serial Composition and Atonality.* University of California Press, Berkeley, CA, 1962.

J. Rahn. *Basic Atonal Theory.* Longman, New York, 1980.

BIBLIOGRAPHY

Chapter 5

B. Polster. *The Mathematics of Juggling.* Springer-Verlag, New York, 2003. [Chapter 6 is devoted to change-ringing and can pretty much be read independently of the rest of the book.]

J. Snowdon and W. Snowdon. *Diagrams.* Christopher Groome, Burton Latimer, England, 1978.

A. White. Fabian Stedman: The First Group Theorist? *American Mathematical Monthly* **103**, 771–778, 1996.

Chapter 6

F. P. Brooks, A. L. Hopkins, P. G. Neumann, and W. V. Wright. An experiment in musical composition. *IRE Transactions on Electronic Computers* **6**, 175–182, 1957. [IRE is the old name of IEEE.]

D. Cope. *Experiments in Musical Intelligence.* A-R Editions, Madison, WI, 1996.

D. Cope. *Virtual Music.* MIT Press, Cambridge, MA, 2001. [Includes an audio CD with musical examples.]

S. Hedges. Dice music in the eighteenth century. *Music and Letters* **59**, 180–187, 1978.

L. Hiller and L. Isaacson. *Experimental Music.* McGraw-Hill, New York, 1959. [Describes the Illiac Suite.]

J. G. Kemeny, J. L. Snell, and G. L. Thompson. *Introduction to Finite Mathematics.* http://math.dartmouth.edu/~doyle/docs/finite/cover/cover.html. [A classic textbook, now out of print, but downloadable for free.]

T. H. O'Beirne. 940,369,969,152 Dice-music trios. *Musical Times* **109**, 911–913, 1968. [Contains the Haydn(?) minuet.]

L. Ratner. Ars combinatoria: Chance and choice in eighteenth-century music. In H. C. Robbins Landon and R. Chapman (editors), *Studies in Eighteenth-Century Music*. Oxford University Press, Oxford, 1970.

Chapter 7

R. F. Voss and J. Clarke. "$1/f$ noise" in music. *Journal of the Acoustical Society of America* **63**, 258–263, 1978.

M. Gardner. *Fractal Music, Hypercards and More.* . . . W. H. Freeman, New York, 1992.

Chapter 8

L. Copes. Dancing with mathematics: Mathematics related to contra dancing. http://www.edmath.org/copes/contra/.

M. Gardner. *Fractal Music, Hypercards and More.* . . . W. H. Freeman, New York, 1992.

S. Mason and M. Saffle. L-systems, melodies and musical structure. *Leonardo Music Journal* **4**, 31–38, 1994.

P. Prusinkiewicz. Score generation with L-systems. In *Proceedings of the 1986 International Computer Music Conference*. International Computer Music Association, 1986.

E. Tarasti. *Heitor Villa-Lobos: The Life and Works, 1887–1959*. McFarland, Jefferson, NC, 1995.

Chapter 9

G. D. Birkhoff. *Aesthetic Measure*. Harvard University Press, Cambridge, MA, 1933. [Also, see A Mathematical Theory of Aesthetics and other papers in Volume III of Birkhoff's *Collected Papers*, American Mathematical Society, 1950.]

U. Dudley. *Numerology, or, What Pythagoras Wrought*. Mathematical Association of America, Washington, DC, 1997.

BIBLIOGRAPHY

G. Markowsky. Misconceptions about the golden ratio. *College Mathematics Journal* **23**, 2–19, 1992.

J. Putz. The golden section and the piano sonatas of Mozart. *Mathematics Magazine* **64**, 275–282, 1995.

L. Somfai. *Béla Bartók: Composition, Concepts, and Autograph Sources.* University of California Press, Berkeley, CA, 1996.

CONTENTS OF THE CD

The following table tells, for each track of the CD, what is on the track and which page in the text references the track.

CONTENTS OF THE CD

Track 2: Excerpt from the CD "Auditory Demonstrations." Reproduced with permission of A. J. M. Houtsma, T. D. Rossing, and W. M. Wagenaars; Tracks 3–5: Renditions by Erich Neuwirth. Reproduced by permission; Track 14: Two-Voice Invention on a Twelve–Tone Row, by Hanns Jelinek. © 1949 (renewed) Universal Edition A. G., Vienna. All Rights Reserved. Reproduced by permission of European American Music Distributors LLC, U.S. and Canadian agent for Universal Edition A. G., Vienna; Track 15: Concerto for Violin and Orchestra, Op. 36, by Arnold Schoenberg. Copyright © 1939 (renewed) by G. Schirmer, Inc. (ASCAP). International Copyright Secured. All Rights Reserved. Reproduced by Permission. Performed by Graybert Beacham (violin) and Leon Harkleroad (piano); Track 16: Three Compositions for Piano, by Milton Babbitt. © Copyright 1957 by Boelke-Bomart, Inc. Reproduced by permission; Track 17: Reproduced with permission of North American Guild of Change Ringers; Tracks 21–24: Illiac Suite (excerpts), by Lejaren Hiller and Leonard Isaacson. Reproduced with permission of Theodore Presser Company. Performed by the University of Illinois Composition String Quartet: William Mullen (violin), David Rosenboom (violin), Theodore Lucas (viola), and Lee Duckles (cello); Tracks 1, 6–16, 18–20, and 25–27 performed by Leon Harkleroad.

ILLUSTRATION CREDITS

Chapter 2, Figure 1; Chapter 7, Figures 6, 7 and 8. © 1978 The Acoustical Society of America. Reprinted by permission of the American Institute of Physics.

Chapter 2, Figures 2 and 10. © 1992 From *Exploring Music* by Charles Taylor. Reproduced by permission of Routledge/Taylor & Francis Group, LLC.

Chapter 4, Figure 18. *Two-Voice Invention on a Twelve-Tone Row*, by Hanns Jelinek. © 1949 (renewed) Universal Edition A. G., Vienna. All Rights Reserved. Reproduced by permission of European American Music Distributors LLC, U.S., and Canadian agent for Universal Edition A. G., Vienna.

Chapter 4, Figure 19. Concerto for Violin and Orchestra, Op. 36, by Arnold Schoenberg. © 1939 (renewed) by G. Schirmer, Inc. (ASCAP). International Copyright Secured. All Rights Reserved. Reprinted by permission.

Chapter 4, Figure 20. © 1957 by Boelke-Bomart, Inc. Reprinted by permission.

Chapter 5, Figure 2. Reprinted with permission of Brian Zook and Thomas Miller, Philadelphia Guild of Change Ringers.

INDEX

INDEX